Springer Series in

CHEMICAL PHYSICS 90

Springer Series in

CHEMICAL PHYSICS

The purpose of this series is to provide comprehensive up-to-date monographs in both well established disciplines and emerging research areas within the broad fields of chemical physics and physical chemistry. The books deal with both fundamental science and applications, and may have either a theoretical or an experimental emphasis. They are aimed primarily at researchers and graduate students in chemical physics and related fields.

75 **Basic Principles in Applied Catalysis**
By M. Baerns

76 **The Chemical Bond**
A Fundamental Quantum-Mechanical Picture
By T. Shida

77 **Heterogeneous Kinetics**
Theory of Ziegler-Natta-Kaminsky Polymerization
By T. Keii

78 **Nuclear Fusion Research**
Understanding Plasma-Surface Interactions
Editors: R.E.H. Clark and D.H. Reiter

79 **Ultrafast Phenomena XIV**
Editors: T. Kobayashi, T. Okada, T. Kobayashi, K.A. Nelson, and S. De Silvestri

80 **X-Ray Diffraction by Macromolecules**
By N. Kasai and M. Kakudo

81 **Advanced Time-Correlated Single Photon Counting Techniques**
By W. Becker

82 **Transport Coefficients of Fluids**
By B.C. Eu

83 **Quantum Dynamics of Complex Molecular Systems**
Editors: D.A. Micha and I. Burghardt

84 **Progress in Ultrafast Intense Laser Science I**
Editors: K. Yamanouchi, S.L. Chin, P. Agostini, and G. Ferrante

85 **Quantum Dynamics Intense Laser Science II**
Editors: K. Yamanouchi, S.L. Chin, P. Agostini, and G. Ferrante

86 **Free Energy Calculations**
Theory and Applications in Chemistry and Biology
Editors: Ch. Chipot and A. Pohorille

87 **Analysis and Control of Ultrafast Photoinduced Reactions**
Editors: O. Kühn and L. Wöste

88 **Ultrafast Phenomena XV**
Editors: P. Corkum, D. Jonas, D. Miller, and A.M. Weiner

89 **Progress in Ultrafast Intense Laser Science III**
Editors: K. Yamanouchi, S.L. Chin, P. Agostini, and F. Ferrante

90 **Thermodynamics and Fluctuations far from Equilibrium**
By J. Ross

John Ross

Thermodynamics and Fluctuations far from Equilibrium

With a Contribution by R.S. Berry

With 74 Figures

Professor Dr. John Ross
Stanford University, Department of Chemistry
333, Campus Drive, Stanford, CA 94305-5080, USA
E-Mail: john.ross@stanford.edu

Contributor:
Professor Dr. R.S. Berry
University of Chicago, Department of Chemistry and the James Franck Institute
5735, South Ellis Avenue, Chicago, IL 60637, USA
E-Mail: berry@uchicago.edu

ISSN 0172-6218
ISBN 978-3-642-09395-1 e-ISBN 978-3-540-74555-6

Springer is a part of Springer Science+Business Media.

springer.com

Softcover reprint of the hardcover 1st edition 2008

Cover design: eStudio Calamar Steinen

This book is dedicated to
My students
My coworkers
My family

Preface

Thermodynamics is one of the foundations of science. The subject has been developed for systems at equilibrium for the past 150 years. The story is different for systems not at equilibrium, either time-dependent systems or systems in non-equilibrium stationary states; here much less has been done, even though the need for this subject has much wider applicability. We have been interested in, and studied, systems far from equilibrium for 40 years and present here some aspects of theory and experiments on three topics:

Part I deals with formulation of thermodynamics of systems far from equilibrium, including connections to fluctuations, with applications to non-equilibrium stationary states and approaches to such states, systems with multiple stationary states, reaction diffusion systems, transport properties, and electrochemical systems. Experiments to substantiate the formulation are also given.

In Part II, dissipation and efficiency in autonomous and externally forced reactions, including several biochemical systems, are explained.

Part III explains stochastic theory and fluctuations in systems far from equilibrium, fluctuation–dissipation relations, including disordered systems.

We concentrate on a coherent presentation of our work and make connections to related or alternative approaches by other investigators. There is no attempt of a literature survey of this field.

We hope that this book will help and interest chemists, physicists, biochemists, and chemical and mechanical engineers. Sooner or later, we expect this book to be introduced into graduate studies and then into undergraduate studies, and hope that the book will serve the purpose.

My gratitude goes to the two contributors of this book: Prof. R. Stephen Berry for contributing Chap. 14 and for reading and commenting on much of the book, and Dr. Marcel O. Vlad for discussing over years many parts of the book.

Stanford, CA *John Ross*
January 2008

Contents

Part I

Thermodynamics and Fluctuations Far from Equilibrium

1

Introduction to Part I

Thermodynamics is an essential part of many fields of science: chemistry, biology, biotechnology, physics, cosmology, all fields of engineering, earth science, among others. Thermodynamics of systems at equilibrium has been developed for more than one hundred years: the presentation of Willard Gibbs [1] is precise, authoritative and erudite; it has been followed by numerous books on this subject [2–5], and we assume that the reader has at least an elementary knowledge of this field and basic chemical kinetics.

In many instances in all these disciplines in science and engineering, there is a need of understanding systems far from equilibrium, for one example systems in vivo.

In this book we offer a coherent presentation of thermodynamics far from, and near to, equilibrium. We establish a thermodynamics of irreversible processes far from and near to equilibrium, including chemical reactions, transport properties, energy transfer processes and electrochemical systems. The focus is on processes proceeding to, and in non-equilibrium stationary states; in systems with multiple stationary states; and in issues of relative stability of multiple stationary states. We seek and find state functions, dependent on the irreversible processes, with simple physical interpretations and present methods for their measurements that yield the work available from these processes. The emphasis is on the development of a theory based on variables that can be measured in experiments to test the theory. The state functions of the theory become identical to the well-known state functions of equilibrium thermodynamics when the processes approach the equilibrium state. The range of interest is put in the form of a series of questions at the end of this chapter.

Much of the material is taken from our research over the last 30 years. We shall reference related work by other investigators, but the book is not intended as a review. The field is vast, even for just chemistry.

1.1 Some Basic Concepts and Definitions

We consider a macroscopic system in a state, not at equilibrium, specified by a given temperature and pressure, and given Gibbs free energy. For a spontaneous, naturally occurring reaction proceeding towards equilibrium at constant temperature T, and constant external pressure p, a necessary and sufficient condition for the Gibbs free energy change of the reaction is

$$\Delta G \leq 0. \tag{1.1}$$

For a reaction at equilibrium, a reversible process, the necessary and sufficient condition is

$$\Delta G = 0. \tag{1.2}$$

Another important property of ΔG is that it is a Lyapunov function in that it obeys (1.1) and (1.3)

$$\frac{\mathrm{d}\Delta G}{\mathrm{d}t} \geq 0. \tag{1.3}$$

where t is time, until equilibrium is reached. Then (1.2) and (1.4) hold

$$\frac{\mathrm{d}\Delta G}{\mathrm{d}t} = 0. \tag{1.4}$$

A Lyapunov function indicates the direction of motion of the system in time (there will be more on Lyapunov functions later).

An essential task of thermodynamics is the prediction of the (maximum) work that a system can do, such as a chemical reaction; for systems at constant temperature and pressure the change in the Gibbs free energy gives that maximum work other than pressure–volume work.

Systems not at equilibrium may be in a transient state proceeding towards equilibrium, or in a transient state proceeding to a non-equilibrium stationary state, or in yet more complicated dynamical states such as periodic oscillations of chemical species (limit cycles) or chaos. The first two conditions are well explained with an example: consider the reaction sequence

$$A \Leftrightarrow X \Leftrightarrow B, \tag{1.5}$$

in which k_1 and k_2 are the forward and backward rate coefficients for the first ($A \Leftrightarrow X$) reaction and k_3 and k_4 are the corresponding rates for the second reaction. In this sequence A is the reactant, X the intermediate, and B the product. For simplicity let the chemical species be ideal gases, and let the reactions occur in the schematic apparatus, Fig. 1.1, at constant temperature.

We could equally well choose concentrations of chemical species in ideal solutions, and shall do so later. Now we treat several cases:

1. The pressures p_{A} and p_{B} are set at values such that their ratio equals the equilibrium constant K

$$\frac{p_B}{p_A} = K. \tag{1.6}$$

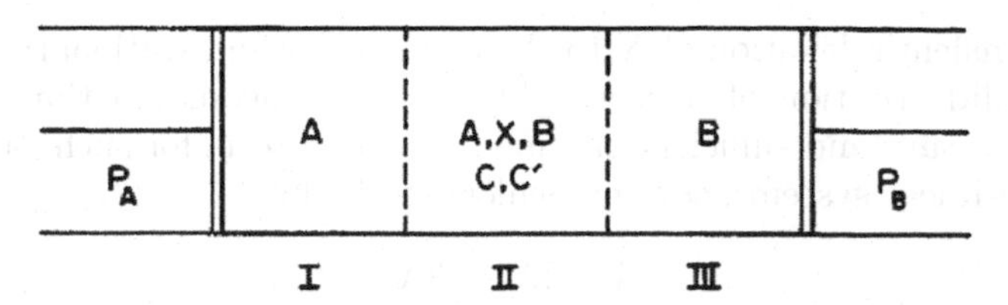

Fig. 1.1. Schematic diagram of two-piston model. The reaction compartment (II) is separated from a reservoir of species A (I) by a membrane permeable only to A and from a reservoir of species B (III) by a membrane permeable only to B. The pressures of A and B are held fixed by constant external forces on the pistons. Catalysts C and C' are required for the reactions to occur at appreciable rates and are contained only in region II

If the whole system is at equilibrium then the concentration of X is

$$X^{\mathrm{eq}} = \frac{k_1}{k_2} A = \frac{k_4}{k_3} B, \tag{1.7}$$

and K can be expressed in terms of the ratio of rate coefficients

$$K = \frac{k_1 k_3}{k_2 k_4}. \tag{1.8}$$

At equilibrium $\Delta G = 0$, or in terms of the chemical potentials $\mu_{\mathrm{A}} = \mu_{\mathrm{B}} = \mu_{\mathrm{X}}$.

2. The pressures of A and B are set as in case 1. If the initial concentration of X is larger than X^{eq} then a transient decrease of X occurs until $X = X^{\mathrm{eq}}$. For the transient process of the system towards equilibrium ΔG of the system is negative, $\Delta G < 0$.
3. The pressures of A and B are set such that

$$\frac{p_B}{p_A} < K. \tag{1.9}$$

Then for a given initial value of p_{X} a transient change in p_{X} occurs until a non-equilibrium state is reached. The pressure at that stationary state must be determined from the kinetic equations of the system. For mass action kinetics the deterministic kinetic equations (neglect of fluctuations in the pressures or concentrations) are

$$\frac{\mathrm{d}p_X}{\mathrm{d}t} = k_1 p_A + k_4 p_B - p_X (k_2 + k_3). \tag{1.10}$$

Hence at the non-equilibrium stationary state, where by definition $\frac{\mathrm{d}p_X}{\mathrm{d}t} = 0$, we have for the pressure of X at that state

$$p_X{}^{\mathrm{ss}} = \frac{k_1 p_A + k_4 p_B}{k_2 + k_3}. \tag{1.11}$$

For the transient relaxation of X to the non-equilibrium stationary state ΔG is not a valid criterion of irreversibility or spontaneous reaction. We shall develop necessary and sufficient thermodynamic criteria for such cases.

For non-linear systems, say the Schlögl model [6]

$$A + 2X \Leftrightarrow 3X \tag{1.12}$$

$$X \Leftrightarrow B \tag{1.13}$$

with the rate coefficients k_1 and k_2 for the forward and reverse reaction in (1.12), and k_3 and k_4 in (1.13), there exists the possibility of multiple stationary states for given constraints of the pressures p_A and p_B. The kinetic equation for p_X is

$$\frac{dp_X}{dt} = k_1 p_A p_X^2 + k_4 p_B - \left(k_2 p_X^3 + k_3 p_X\right), \tag{1.14}$$

which is cubic in p_X and hence may have three stationary states (right hand side of (1.14) equals zero) Fig. 1.2.

The region of multiple stationary states extends for the pump parameter (equal to p_A/p_B) from F_1 to F_3; the line segments with positive slope, marked α and γ, are branches of stable stationary states, the line segment with negative slope, marked β, is a branch of unstable stationary states. A system started at an unstable stationary state will proceed to a stable stationary state along

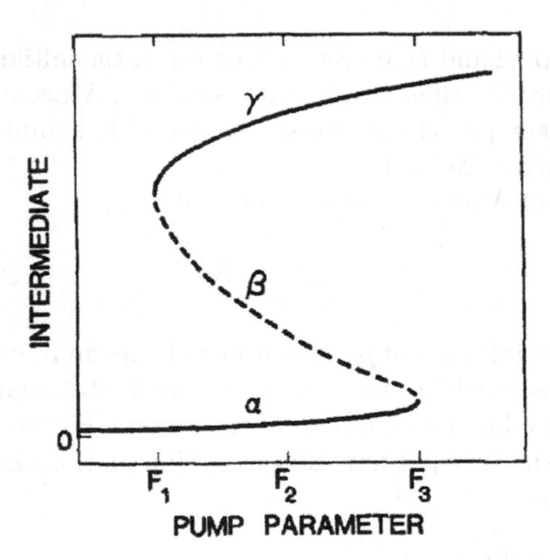

Fig. 1.2. Stationary states of the Schlögl model with fixed reactant and products pressures. Plot of the pressure of the intermediate p_X vs. the pump parameter (p_A/p_B). The branches of stable stationary states are labeled α and γ and the branch of unstable stationary states is labeled β. The marginal stability points are at F_1 and F_3 and the system has two stable stationary states between these limits. The equistability point of the two stable stationary states is at F_2

a deterministic trajectory. The so-called marginal stability points are at F_1 and F_3. For a dcterministic system, for which fluctuations are very small, transitions from one stable branch to the other occur at the marginal stability points. If fluctuations are taken into account then the point of equistability is at F_2, where the probability of transition from one stable branch to the other equals the probability of the reverse transition.

An examples of such systems in the gas phase is the illuminated reaction $S_2O_6F_2 = 2SO_3F$, [7]. An example of multiple stationary states in a liquid phase (water) is the iodate-arseneous acid reaction, [8]. Both examples can be analyzed effectively as one-variable systems.

1.2 Elementary Thermodynamics and Kinetics

Let us consider J coupled chemical reactions with L species proceeding to equilibrium, and let the stoichiometry of the jth reaction, with $1 \leq j \leq J$, be

$$\sum_{l=1}^{L} \nu_{jl} X_l = 0. \tag{1.15}$$

The stoichiometric coefficient ν_{ji} is negative for a reactant, zero for a catalyst and positive for a product. We introduce progress variables ξ_j for each of the j reactions

$$dn_l = \sum_{j-1}^{J} \nu_{jl} d\xi_j \tag{1.16}$$

where n_i denotes number of moles of species i, and the affinities A_j [9]

$$A_j = -\sum_{l=1}^{L} \nu_{jl} \mu_l, \tag{1.17}$$

expressed in terms of the chemical potentials μ_l. (The introduction of chemical potentials in chemical kinetics requires the assumption of local equilibrium, which is discussed in Chap. 2.) With (1.17) we write the differential change in Gibbs free energy for the reactions

$$\mathrm{d}\Delta G = -\sum_{j=1}^{J} A_j \mathrm{d}\xi_j \tag{1.18}$$

For the jth reaction the kinetics can be written

$$\mathrm{d}\xi_j/\mathrm{d}t = t_j^+ - t_j^- (1 \leqslant j \leqslant J), \tag{1.19}$$

where t_j^+, t_k^- are the reaction fluxes for this reaction step in the forward and reverse direction, respectively. Hence the affinities may be rewritten

$$A_j = RT \ln(t_j^+/t_j^-), \tag{1.20}$$

which is easily obtained for any elementary reaction by writing out the t_j^+/t_j^- in terms of concentrations and the introductions of chemical potentials, (2.4). The time rate of change of the Gibbs free energy is

$$\frac{\mathrm{d}\Delta G}{\mathrm{d}t} = -\sum_{j=1}^{J} A_j \frac{\mathrm{d}\xi_j}{\mathrm{d}t} = -\sum_{j=1}^{J} \left[RT \ln \left(t_j^+/t_j^- \right) \right] \left(t_j^+ - t_j^- \right) \tag{1.21}$$

in which each term on the *rhs* is a product of the affinity of a given reaction times the rate of that reaction. The rate of change of ΔG is negative for every term until equilibrium is reached when ΔG of the reaction is zero. Hence ΔG is a Liapunov function and provides an evolution criterion for the kinetics of the system. The form of (1.21) is the same as that of Boltzmann's H theorem for the increase in entropy during an irreversible process in an isolated system [10].

For an isothermal system we have

$$\mathrm{d}G = \mathrm{d}H - T\mathrm{d}S, \tag{1.22}$$

and hence

$$\frac{\mathrm{d}G}{\mathrm{d}t} = \frac{\mathrm{d}H}{\mathrm{d}t} - T\frac{\mathrm{d}S}{\mathrm{d}t}. \tag{1.23}$$

At constant concentration (chemical potential), and hence pressure for each of the reservoirs we have the relation

$$\frac{\mathrm{d}H}{\mathrm{d}t} = -T\frac{\mathrm{d}S_{\mathrm{rev}}}{\mathrm{d}t}, \tag{1.24}$$

where $\mathrm{d}S_{\mathrm{rev}}$ is the differential change in entropy of the surroundings due to (reversible) passage of heat from the system to the surroundings. Hence we may write

$$\frac{\mathrm{d}G}{\mathrm{d}t} = -T\left[\frac{\mathrm{d}S}{\mathrm{d}t} + \frac{\mathrm{d}S_{\mathrm{rev}}}{\mathrm{d}t}\right] = -T\frac{\mathrm{d}S_{\mathrm{univ}}}{\mathrm{d}t}, \tag{1.25}$$

that is the product of T and the total rate of entropy production in the universe is the dissipation.

For a generalization of the model reaction, (1.12, 1.13), we write

$$A + (r-1)X \underset{k_2}{\overset{k_1}{\rightleftarrows}} rX,$$

$$(s-1)X + B \underset{k_3}{\overset{k_4}{\rightleftarrows}} sX.$$

for which the variation in time of the intermediate species X is

$$\mathrm{d}p_X/\mathrm{d}t = k_1 p_A p_x^{r-1} - k_2 p_X^r - k_3 p_X^s + k_4 p_B p_X^{s-1}. \tag{1.26}$$

The stability of the stationary states of the system described by this equation can be obtained by linearizing (1.26) around each such state [11]. The stability criteria so obtained are

$$\begin{aligned}
&\mathrm{d}p_X/\mathrm{d}t = 0 \quad \text{at each steady state,}\\
&\mathrm{d}(\mathrm{d}p_X/\mathrm{d}t)/\mathrm{d}p_X < 0 \quad \text{at each stable steady-state,}\\
&\mathrm{d}(\mathrm{d}p_X/\mathrm{d}t)/\mathrm{d}p_X > 0 \quad \text{at each unstable steady-state,}\\
&\mathrm{d}(\mathrm{d}p_X/\mathrm{d}t)/\mathrm{d}p_X = 0 \quad \text{at each marginally stable steady-state,}\\
&\quad \text{and}\\
&\mathrm{d}(\mathrm{d}p_X/\mathrm{d}t)/\mathrm{d}p_X = \mathrm{d}^2(\mathrm{d}p_X/\mathrm{d}t)/\mathrm{d}p_X^2 = 0 \quad \text{at each critically stable}\\
&\quad \text{steady-state.}
\end{aligned} \tag{1.27}$$

At a critically steady (stationary) state the left and right marginal stability points coincide.

In the next few chapters, we shall formulate these kinetic criteria in terms of thermodynamic concepts.

Several important issues need to be addressed in non-equilibrium thermodynamics:

What are the thermodynamic functions that describe the approach of such systems to a non-equilibrium stationary state, both the approach of each intermediate species and the reaction as a whole?

How much work can be obtained in the surroundings of a system relaxing to a stable stationary state?

How much work is necessary to move a system in a stable stationary state away from that state?

What are the thermodynamic forces, conjugate fluxes and applicable extremum conditions for processes proceeding to or from non-equilibrium stationary states? What is the dissipation for these processes?

What are the suitable thermodynamic Lyapunov functions (evolution criteria)?

What are the relations of these thermodynamic functions, if any, to ΔG?

What are the relations of these thermodynamic functions to the work that a system can do in its approach to a stable stationary state?

What are the necessary and sufficient thermodynamic criteria of stability of the various branches of stationary states?

What are the thermodynamic criteria of relative stability in the region where there exist two or more branches of stable stationary states? What are the necessary and sufficient thermodynamic criteria of equistability of two stable stationary states?

What are the thermodynamic conditions of marginal stability?

What are interesting and useful properties of the dissipation?

We shall provide answers to some of these questions in Chap. 2 for one variable systems, based on a deterministic analysis. In later chapters, we discuss relevant experiments and compare with the theory.

Then we address these same questions in Chap. 3 for multivariable systems, with two or more intermediates. Now our approach takes inherent fluctuations fully into account and we find a state function (analogous to ΔG) that satisfies the stated requirements. We also present a deterministic analysis of multivariable systems in Chap. 4 and compare the approach and the results with the fluctuational analysis. In Chap. 5 we turn to the study of reaction-diffusion systems and the issue of relative stability of multiple stationary states. The same issue is addressed in Chap. 6 on the basis of fluctuations, and in Chap. 7 we present experiments on relative stability.

The thermodynamics of transport properties, diffusion, thermal conduction and viscous flow is taken up in Chap. 8, and non-ideal systems are treated in Chap. 9. Electrochemcial experiments in chemical systems in stationary states far from equilibrium are presented in Chap. 10, and the theory for such measurements in Chap. 11 in which we show the determination of the introduced thermodynamic and stochastic potentials from macroscopic measurements.

Part I concludes with the analysis of dissipation in irreversible processes both near and far from equilibrium, Chap. 12.

There is a substantial literature on this and related subjects that we shall cite and comment on briefly throughout the book.

Acknowledgement. A part of the presentation in this chapter is taken from ref. [12].

References

1. J.W. Gibbs, *The Collected Works of J.W. Gibbs*, vol. I. Thermodynamics (Yale University Press, 1948)
2. A.A. Noyes, M.S. Sherril, *A Course of Study in Chemical Principles* (MacMillan, New York, 1938)
3. G.N. Lewis, M. Randall, *Thermodynamics*, 2nd ed., revised by K.S. Pitzer, L. Brewer, (McGraw-Hill, New York, 1961)
4. J.G. Kirkwood, I. Oppenheim, *Chemical Thermodynamics* (McGraw-Hill, New York, 1961)
5. R.S. Berry, S.A. Rice, J. Ross, *Physical Chemistry*, 2nd edn. (Oxford University Press, 2000)
6. F. Schlögl, Z. Phys. **248**, 446–458 (1971)
7. E.C. Zimmermann, J. Ross, J. Chem. Phys. **80**, 720–729 (1984)
8. N. Ganapathisubramanian, K. Showalter, J. Chem. Phys. **80**, 4177–4184 (1984)
9. G. Nicolis, I. Prigogine, *Self-Organization in Nonequilibrium Systems* (Wiley, New York, 1977)
10. R.C. Tolman, *The Principles of Statistical Mechanics* (Oxford University Press, London, 1938)
11. L. Cesari, *Asymptotic Behavior and Stability Problems in Ordinary Differential Equations*, 3rd edn. (Springer, Berlin Heidelberg New York, 1980)
12. J. Ross, K.L.C. Hunt, P.M. Hunt, J. Chem. Phys. **88**, 2719–2729 (1988)

2

Thermodynamics Far from Equilibrium: Linear and Nonlinear One-Variable Systems

2.1 Linear One-Variable Systems

We begin as simply as possible, with a linear system, (1.5), repeated here

$$A \Leftrightarrow X \Leftrightarrow B, \tag{2.1}$$

with rate coefficients k_1 and k_2 for the rate coefficients in the forward and reverse reaction of the first reaction, and similarly k_3 and k_4 for the second reaction. The deterministic rate equation is (1.10), rewritten here in a slightly different form,

$$\frac{\mathrm{d}p_X}{\mathrm{d}t} = (k_1 p_A + k_4 p_B) - (k_2 + k_3)\, p_X \tag{2.2}$$

for isothermal ideal gases; the pressures of A and B are held constant in an apparatus as in Fig. 1.1 of Chap. 1. We denote the first term on the rhs of (2.2) by t_X^+ and the second term by t_X^- [1]. The pressure of p_X at the stationary state, with the rhs of (2.2) set to zero, is

$$\frac{p_X^{\mathrm{s}}}{p_X} = \frac{t_X^+}{t_X^-} = \frac{t_X^{+\mathrm{s}}}{t_X^-}. \tag{2.3}$$

since t_X^+ is a constant.

Now we need an important hypothesis, that of *local equilibrium*. It is assumed that at each time there exists a temperature, a pressure, and a chemical potential for each chemical species. These quantities are established on time scales short compared with changes in pressure, or concentration, of chemical species due to chemical reaction. Although collisions leading to chemical reactions may perturb, for example, the equilibrium distribution of molecular velocities, that perturbation is generally small and decays in 10–30 ns, a time scale short compared with ranges of reaction rates of micro seconds and longer. There are many examples that fit this hypothesis well [2]. (A phenomenological approach beyond local equilibrium is given in the field of extended irreversible thermodynamics [3, 4], which we do not discuss here.) We thus write for the chemical potential

$$\mu_X = \mu_X^0 + RT \ln p_X \tag{2.4}$$

where μ_X^0 is the chemical potential in the standard state. Hence we have

$$\mu_X - \mu_X^{\mathrm{s}} = -RT \ln \frac{t_X^+}{t_X^-}. \tag{2.5}$$

We define a thermodynamic state function ϕ [1]

$$\phi\left(p_X\right) = V_{\mathrm{II}} \int \left(\mu_X - \mu_X^{\mathrm{s}}\right) dp_X \tag{2.6}$$

where V_{II} is a volume shown in Fig. 1.1 of Chap. 1. This function has many important properties. At the stationary state of this system ϕ is zero. If we start at the stationary state and increase p_X then $\mathrm{d}p_X \geq 0$ and the integrand is larger than zero. Hence ϕ is positive. Similarly, if we start at the stationary state and decrease p_X then $\mathrm{d}p_X$ and the integrand are both negative and ϕ is positive. Hence ϕ is an extremum at the stable stationary state, a minimum.

Before discussing further properties of this state function, we can proceed to nonlinear one-variable systems, which also have only one intermediate.

2.2 Nonlinear One-Variable Systems

We write a model stoichiometric equation

$$\begin{aligned} A + (r-1)\, X &\underset{k_2}{\overset{k_1}{\rightleftarrows}} rX, \\ (s-1)\, X + B &\underset{k_3}{\overset{k_4}{\rightleftarrows}} sX. \end{aligned} \tag{2.7}$$

and imagine this reaction occurring in the apparatus, Fig. 1.1 of Chap. 1. Since this isothermal systems has chambers I and III at constant pressure and chamber II at constant volume the proper thermodynamic function for the entire system is a linear sum of Gibbs free energies for I and III and the Helmholtz free energy for II. If in (2.7) $s = 1$ and $r = 1$ then we have the linear model (2.1). If we set $r = 3$ and $s = 1$ then we have the Schlögl model, (1.12, 1.13). We shall use the results obtained above for the linear model to develop results for the Schlögl model. The deterministic kinetic equation for the Schlögl model was given in (1.14) and is repeated here

$$\frac{\mathrm{d}p_X}{\mathrm{d}t} = k_1 p_A p_X^2 + k_4 p_B - \left(k_2 p_X^3 + k_3 p_X\right). \tag{2.8}$$

The first two positive terms on the rhs of (2.8) are again given the symbol t_X^+ and the two negative terms the symbol t_X^-; their ratio is

$$\frac{t_X^+}{t_X^-} = \frac{k_1 p_A p_X^2 + k_4 p_B}{k_2 p_X^3 + k_3 p_X}, \tag{2.9}$$

which we use to define the quantity p_X^*

$$\frac{t_X^+}{t_X^-} = \frac{p_X^*}{p_X}. \tag{2.10}$$

Hence p_X^* is

$$p_X^* = \frac{k_1 p_A p_X^2 + k_4 p_B}{k_2 p_X^2 + k_3}. \tag{2.11}$$

The quantity p_X^* is the pressure in a reference state for which (2.10) holds.

If we compare (2.3) with (2.10) we see the similarity obtained by defining p_X^*. We gain some insight by comparing the linear model with the Schlögl model in the following way: assign the same value of p_A to each, the same value of p_B to each, and similarly for T, V_I, V_{II}, V_{III}, the equilibrium constant for the $A \Leftrightarrow X$ reaction and that for the $B \Leftrightarrow X$ reaction. Then the two model systems are 'instantaneously thermodynamically equivalent.' If furthermore t_X^+ has the same value in the two systems at each point in time, and the same for t_X^-, the two systems are 'instantaneously kinetically indistinguishable.' Hence following (2.5 and 2.6) we may write

$$\mu_X - \mu_X^* = RT \ln \frac{p_X}{p_X^*} = -RT \ln \frac{t_X^+}{t_X^-} \tag{2.12}$$

and for our chosen thermodynamic function

$$\phi^*(p_X) = \int (\mu_X - \mu_X^*) \, \mathrm{d}p_X. \tag{2.13}$$

In the instantaneously indistinguishable linear system $p_X{}^*$ denotes the pressure of X in the stationary state. The function in (2.13) is an *excess work*, the work of moving the system from a stable stationary state to an arbitrary value p_X compared with the work of moving the system from the stationary state of the instantaneous indistinguishable linear system to p_X.

The integrand in (2.13) is a species-specific activity, which plays a fundamental role, as we now show.

The integrand in (2.13) is a state function and so is ϕ^*; as before, ϕ^* is an extremum at the stable stationary state, a minimum. We come to that from

$$\begin{aligned}(d(\mu_x - \mu_x^*)/dp_x)|_{ss} &= -RT\left[(dt_x^+/dp_x)_{ss} - (dt_x^-/dp_x)_{ss}\right]/(t_x^+|_{ss}) \\ &= -V_{II}\left(t_X^+|_{ss}\right)^{-1}\left[d(dp_x/dt)/dp_x\right]|_{ss}\end{aligned} \tag{2.14}$$

and (1.24), so that we have the following necessary and sufficient conditions for the species-specific activity (the driving force for species X)

$$\begin{aligned}
&\mu_x - \mu_x^* = 0 \quad \text{at each steady-state,}\\
&\mathrm{d}\left(\mu_x - \mu_x^*\right)/\mathrm{d}p_x > 0 \quad \text{at each stable steady-state,}\\
&\mathrm{d}\left(\mu_x - \mu_x^*\right)/\mathrm{d}p_x < 0 \quad \text{at each unstable steady-state, and}\\
&\mathrm{d}\left(\mu_x - \mu_x^*\right)/\mathrm{d}p_x = 0 \quad \text{at each marginally sable steady-state.}
\end{aligned} \tag{2.15}$$

In addition we have

$$\mathrm{d}\left(\mu_x - \mu_x^*\right)/\mathrm{d}p_x = \mathrm{d}^2\left(\mu_x - \mu_x^*\right)/\mathrm{d}p_x^2 = 0 \quad \text{at each ciritically stable steady-state.} \tag{2.16}$$

It is useful to restate these results in terms necessary and sufficient conditions for the state function $\phi^*(p_X)$, (2.13):

$$\frac{\mathrm{d}\phi^*}{\mathrm{d}p_X} = 0 \quad \text{at each stationary state} \tag{2.17}$$

$$\frac{\mathrm{d}^2\phi^*}{\mathrm{d}p_X^2} \geq 0 \quad \text{at each stable stationary state with the equality sign holding at marginal stability} \tag{2.18}$$

$$\frac{\mathrm{d}^2\phi^*}{\mathrm{d}p_X^2} \leq 0 \quad \text{at each unstable stationary state with the equality sign holding at marginal stability} \tag{2.19}$$

Hence (2.17, 2.18) are necessary and sufficient conditions for the existence and stability of nonequilibrium stationary states.

There are more conditions to be added after developing the connection of the thermodynamic theory to the stochastic theory.

It may seem strange that in (2.12) the chemical potential difference on the lhs is related to the logarithm of a ratio of fluxes and each flux consists of two additive terms. We can find an interpreation by comparison with a single reaction, that of $A + B = C + D$. We can write the flux in the forward direction

$$t^+ = k_f V\,[A]\,[B] = V\,[A]\,[B]\,\bar{v}_{AB}\bar{\sigma}_{AB}, \tag{2.20}$$

where the brackets indicate concentrations of species, V is the reaction volume, $\bar{v}_{\mathrm{AB}}$ is the average relative speed of A and B, and $\bar{\sigma}_{\mathrm{AB}}$ is the reaction cross section, averaged with a weighting of the relative speed. Hence the term $k_{\mathrm{f}}V$ [A][B] is the flux of A and B to form C and D, and k_{f} [C][D] is the flux pf products to form reactants. The chemical potential difference between the products and reactants is the driving force toward equilibrium and is proportional to the logarithm of the ratio of the fluxes in the forward and reverse direction, see (1.20). For the reaction mechanism (2.7), the flux of reactants to form X comes from two sources: the reaction A with X and the reaction B to form X. The total flux is the sum of fluxes from these two sources. Similarly, the flux of removing X has two sources. In all cases these fluxes are indications of the respective escaping tendencies and hence the relation to the chemical potentials. Thus (2.12) connects the lhs, the chemical driving force toward a stable stationary state, to the ratio of sums of fluxes of X, the rhs.

If A and B are chosen such that the ratio of their pressures equals the equilibrium constant then ϕ^* equals ΔG and $p_X^* = p^s$.

2.3 Dissipation

For a spontaneously occurring chemical reaction at constant pressure, p, and temperature, T, the Gibbs free energy change gives the maximum work, other than pV work, that can be obtained from the reaction. For systems at constant V, T it is the Helmholtz free energy change that yields that measure. If no work is done by the reaction then the respective free energy changes are dissipated, lost. For reactions of ideal gases run in the apparatus in Fig. 1.1 in Chap. 1, we can define a hybrid free energy, M,

$$M = \mu_A \left(n_A^{\mathrm{I}} + n_A^{\mathrm{II}}\right) + \mu_B \left(n_B^{\mathrm{II}} + n_B^{\mathrm{III}}\right) + \mu_x n_x^{\mathrm{II}} \\ -RT \left(n_A^{\mathrm{II}} + n_B^{\mathrm{II}} + n_x^{\mathrm{II}}\right). \quad (2.21)$$

The time rate of change of M is

$$\mathrm{d}M/\mathrm{d}t = \mu_A \mathrm{d}n_A^{\mathrm{I}}/\mathrm{d}t + \mu_B \, \mathrm{d}n_B^{\mathrm{III}}/\mathrm{d}t + \mu_x \, \mathrm{d}n_x^{\mathrm{II}}/\mathrm{d}t \quad (2.22)$$

if there is no depletion of the reservoirs I and III. According to conservation of mass we have

$$0 = \mu_x^* \, \mathrm{d}n_A^{\mathrm{I}}/\mathrm{d}t + \mu_x^* \mathrm{d}n_B^{\mathrm{III}}/\mathrm{d}t + \mu_x^* \mathrm{d}n_x^{\mathrm{II}}/\mathrm{d}t, \quad (2.23)$$

and therefore we may write

$$\mathrm{d}M/\mathrm{d}t = (\mu_A - \mu_x^*) \, \mathrm{d}n_A^{\mathrm{I}}/\mathrm{d}t + (\mu_B - \mu_x^*) \, \mathrm{d}n_B^{\mathrm{III}}/\mathrm{d}t \\ + (\mu_x - \mu_x^*) \, \mathrm{d}n_x^{\mathrm{II}}/\mathrm{d}t. \quad (2.24)$$

Hence we write for the dissipation D

$$D = -\mathrm{d}M/\mathrm{d}t = -\mathrm{d}M_{\mathrm{res}}/\mathrm{d}t - \mathrm{d}M_x/\mathrm{d}t \quad (2.25)$$

where the first term on the rhs is the dissipation due to the conversion of A to X at the pressure p_X^* and at the rate $-\frac{dn_A^I}{dt}$ and the conversion of X to B at the same pressure of X and the rate $\frac{\mathrm{d}n_B^{III}}{\mathrm{d}t}$. The second term on the rhs of (2.25) is

$$-\mathrm{d}M_x/\mathrm{d}t = -(\mu_x - \mu_x^*) \, \mathrm{d}n_x^{\mathrm{II}}/\mathrm{d}t \\ = RT \left(t_x^+ - t_x^-\right) \ln \left(t_x^+/t_x^-\right) \\ \equiv D_{\mathrm{x}}. \quad (2.26)$$

From this last equation it is clear that we have for D_X

$$D_x = -\mathrm{d}M_x/\mathrm{d}t \geq 0 \quad \text{for all } p_x, \tag{2.27}$$

regardless of the reaction mechanism; the equality holds only at the stationary state.

As we shall discuss later, the total dissipation D is not an extremum at stationary states in general, but there may be exceptions. D_X is such an extremum and the integral

$$\phi^* = \int (\mu_X - \mu_X^*)\,\mathrm{d}n_X \tag{2.28}$$

is a Lyapunov function in the domain of attraction of each stable stationary state.

The dissipation in a reaction can range from zero, for a reversible reaction, to its maximum of ΔG when no work is done in the surroundings. Hence the dissipation can be taken to be a measure of the efficiency of a reaction in regard to doing work. There is more on this subject in Chap. 12.

2.4 Connection of the Thermodynamic Theory with Stochastic Theory

The deterministic theory of chemical kinetics is formulated in terms of pressures, for gases, or concentrations of species for gases and solutions. These quantities are macroscopic variables and fluctuations of theses variables are neglected in this approach. But fluctuations do occur and one way of treating them is by stochastic theory. This kind of analysis is also called mesoscopic in that it is intermediate between the deterministic theory and that of statistical mechanics. In stochastic theory, one assumes that fluctuations do occur, say in the number of particles of a given species X, that there is a probability distribution $P(X,t)$ for that number of particles at a given time, and that changes in this distribution occur due to chemical reactions. The transitions probabilities of such changes are assumed to be given by macroscopic kinetics. We shall show that the nonequilibrium thermodynamic functions ϕ for linear systems, ϕ^* (for nonlinear systems), the excess work, determines the stationary, time-independent, probability distribution, which leads to a physical interpretation of the connection of the thermodynamic and stochastic theory. At equilibrium, the probability distribution of fluctuations is determined by the Gibbs free energy change at constant T, p, which is the work other than pV work.

We restrict the analysis at first to reaction mechanisms for which the number of molecules of species X changes by ± 1 in each elementary step.

We take the probability distribution to obey the master equation which has been used extensively. For the cubic Schlögl model ((2.7) with $r = 3$, $s = 1$) the master equation is [1,5]

$$\partial P(X,t)/\partial = t^{+}(X-1)P(X-1,t) + t^{-}(X+1)P(X+1,t) - \left[t^{+}(X) + t^{-}(X)\right]P(X,t). \tag{2.29}$$

The first two terms on the rhs yield an increase in X, the last two terms a decrease in X.

The fluxes in this equation are

$$t^{+}(X) = c_1 AX(X-1)/2! + c_4 B,$$
$$t^{-}(X) c_2 X(X-1)(X-2)/3! + c_3 X, \tag{2.30}$$

with the parameters c_i related to the rate coefficients k_i by

$$k_i = V^{m_i - 1}(c_i/n_t!) \quad \text{for} \quad 1 \le i \le 4, \tag{2.31}$$

where m_i is the molecularity of the ith step and n_i the molecularity in X.

From the master equation, we can derive the result that the average concentration, the average number of X in a volume V, obeys the deterministic rate equation in the limit of large numbers of molecules.

The time-independent solution of the master equation is

$$P_s(X) = P_s(0) \prod_{i=1}^{X} \frac{t^{+}(i-1)}{t^{-}(i)}, \tag{2.32}$$

which by retention of only the leading term in the Euler-MacLaurin summation formula reduces to

$$P_s(X) = N \exp\left[\int_1^x \ln \frac{t^{+}(y)}{t^{-}(y)} \mathrm{d}y\right] \tag{2.33}$$

and N is a normalization constant. The connection between the thermodynamic and stochastic theory is established with the use of (2.12) to give

$$P_s(X) = N \exp\left[-\frac{1}{RT}\int^x (\mu_x - \mu_x^*)\,\mathrm{d}X\right], \tag{2.34}$$

The Lyapunov function ϕ^*, (2.13), is both the thermodynamic driving force toward a stable stationary state and determines the stationary probability distribution of the master equation. The stationary distributions (2.33, 2.34) are nonequilibrium analogs of the Einstein relations at equilibrium, which give fluctuations around equilibrium.

There is another interesting connection [1]. We define $P(X_1, t_1;\ X_0, t_0)$ to be the probability density of observing X_1 molecules in V at time t_1 given that there are X_0 molecules at t_0. This function is the solution of the master equation (2.29) for the initial condition

$$P(X, t = t_0) = \delta(X - X_0). \tag{2.35}$$

The probability density can be factored into two terms, [1],

$$P(X_1, t_1; X_0, t_0) = F_1(X_0 \to X_1)\, F_2(X_1, X_0, t_1 - t_0), \tag{2.36}$$

in which the first term on the rhs is independent of the path from X_0 to X_1 and independent of the time interval $(t_1 - t_0)$. To the same approximation with which we obtained (2.33) we can reduce the first term to

$$F_1(X_0 \to X_1) = \exp\left[\left(\frac{1}{2}\right)\int_{x_0}^{x_1} (\ln t^+/t^-)\, \mathrm{d}X\right], \tag{2.37}$$

and find it to be of the same form as the probability distribution (2.33). It contains the irreversible part of the probability density (2.36).

2.5 Relative Stability of Multiple Stationary Stable States

For systems with multiple stable stationary states there arises the issue of relative stability of such states. As in the previous section we treat systems with a single intermediate and stoichiometric changes in X are limited to ± 1.

In regions of multistability the stationary probability distribution is bimodal and is shown in Fig. 2.1 for the cubic Schlögl model.

Stable stationary states are located at maxima, labeled 1 and 3, and unstable stationary states at minima, labeled 2.

Consider now the ratio of the probability density (2.36), for a given transition from X_1 to X_2 to that of the reverse transition

$$\frac{P(X_2, t_2; X_1, t_1)}{P(X_1, t_2; X_2, t_1)} = \exp\left[\int_1^2 \ln \frac{t_x^+}{t_x^-} \mathrm{d}X\right] \tag{2.38}$$

We obtain this equation with the use of (2.36, 2.37), once for the numerator and once for the denominator on the lhs of (2.38), canceling the F_2 terms, and moving the remaining term in the denominator to the numerator. Equistability of two stable stationary states, labeled now 1 and 3 to correspond to Fig. 2.1, is defined by

$$\frac{P(X_3, t; X_1, 0)}{P(X_1, t; X_3, 0)} = 1, \tag{2.39}$$

which with the use of the second and third line of (2.26) we may also express as

$$\int_2^3 D_x\, \mathrm{d}t = \int_2^1 D_x\, \mathrm{d}t. \tag{2.40}$$

The integral of the species-specific dissipation from the unstable stationary state 2 to the stable stationary state 3 equals, at equistability, to the integral

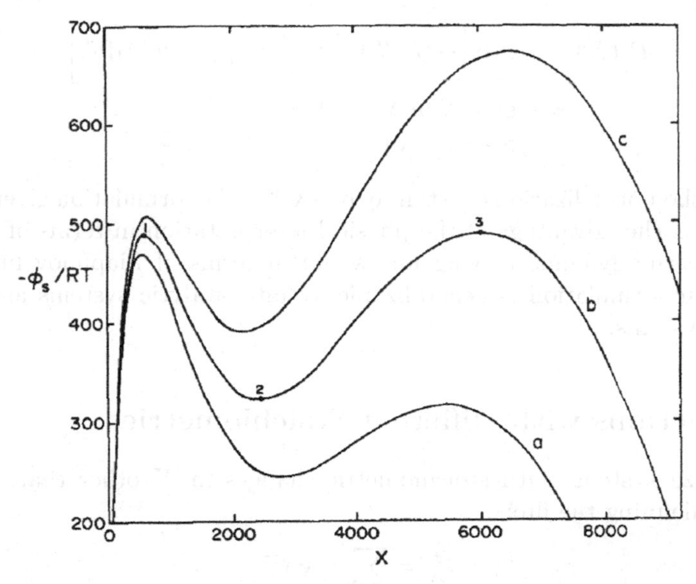

Fig. 2.1. Plot of the integral in (2.34), marked ϕ_s vs. X for the Schlögl model, (1.12, 1.13), with parameters: $c_1 = 3.10^{-10}\,\mathrm{s}^{-1}$; $c_2 = 1.10^{-7}\,\mathrm{s}^{-1}$; $c^3 = 0.33\,\mathrm{s}^{-1}$; $c_4 = 1.5.10^{-4}\mathrm{s}^{-1}$; and $A = B$. For curve (a) $B = 9.8.10^6$; for curve (b) $B = 1.01.10^6$; curve (c) $B = 1.04.10^6$. Curve (b) lies close to the equistability of the stable stationary states 1 and 3; 2 marks the unstable stationary state

of the dissipation from the unstable stationary state 2 to the stable stationary state 1, whereas the integral of the total dissipation for the limits in (2.40) goes to infinity and that of the species-specific dissipation is finite. We can restate (2.40) in terms of the excess work (see the first and third equation of (2.26)

$$\int_2^3 (\mu_\mathrm{X} - \mu_\mathrm{X}^*)\,\mathrm{d}n_\mathrm{X}^\mathrm{II} = \int_2^1 (\mu_\mathrm{X} - \mu_\mathrm{X}^*)\,\mathrm{d}n_\mathrm{X}^\mathrm{II}; \tag{2.41}$$

at equistability the integral of the excess work from 2 to 3 equals the integral of the excess work from 2 to 1. Equations (2.39–2.41) provide necessary and sufficient conditions of equistability of stable stationary states.

The master equation has been investigated for a sequence of unimolecular (nonautocatalystic) reactions based on moment generating functions [6]; these yield Poissonian stationary distribution for single intermediate systems in terms of the number of particles X of species X, with X^{ss} that number in the stationary state

$$P_s\,(X) = \left[(X^{\mathrm{ss}})^{\mathrm{X}}\,/X!\right] \exp\left(-X^{\mathrm{ss}}\right). \tag{2.42}$$

Our results are consistent with (2.41) as can be seen from the use of (2.13) and (2.34), a change of variables to particle numbers X, and the use of Stirling's approximation

$$\begin{aligned} P_s(X) &= \wp \exp\left[-(k_B T)^{-1} \times \int^{n_x} (\mu_x - \mu_x^{ss})\, dn'_x\right] \\ &= \wp \exp(X \ln X^{ss} - X \ln X + X) \\ &= \wp \left[(X^{ss})^x / X!\right]. \end{aligned} \tag{2.43}$$

Here p is the normalization constant $\exp(-X^{ss})$. The formulation given in this chapter has the advantage of the physical interpretation in terms of species-specific thermodynamic driving forces and in terms of Liapunov functions; further our formulation is generalizable to autocatalytic systems and many variable systems.

2.6 Reactions with Different Stoichiometries

We analyze systems with stoichiometric changes in X other than ± 1. We begin by defining the flux

$$t_X^{\pm} = \sum_j |\nu_{jX}| t_j^{\pm} \tag{2.44}$$

and again choose $p^*{}_X$ for any given p_X so that we have

$$\mu_X - \mu_X^* = RT \ln\left(t_X^- / t_X^+\right). \tag{2.45}$$

Let the reactions occur in the apparatus Fig. 1.1 of Chap. 1; then the rate of change of the mixed free energy M is

$$\mathrm{d}M/\mathrm{d}t = \mu_A \mathrm{d}n_A^{\mathrm{I}}/\mathrm{d}t + \mu_B \mathrm{d}n_B^{\mathrm{III}}/\mathrm{d}t + \mu_X \mathrm{d}n_X^{\mathrm{II}}/\mathrm{d}t, \tag{2.46}$$

and we need to consider conservation of mass. For example, for the reaction mechanism

$$A \rightleftarrows X, \quad 2X \rightleftarrows B$$

mass conservation requires

$$0 = \mu_X^* \mathrm{d}n_A^{\mathrm{I}}/\mathrm{d}t + 2\mu_X^* \mathrm{d}n_B^{\mathrm{III}}/\mathrm{d}t + \mu_X^* \mathrm{d}n_X^{\mathrm{II}}/\mathrm{d}t, \tag{2.47}$$

so that we have

$$\begin{aligned} \frac{\mathrm{d}M}{\mathrm{d}t} &= (\mu_A - \mu_X^*)\mathrm{d}n_A^{\mathrm{I}}/\mathrm{d}t + (\mu_B - 2\mu_X^*)\, \mathrm{d}n_B^{\mathrm{III}}/\mathrm{d}t \\ &\quad + (\mu_X - \mu_X^*)\, \mathrm{d}n_X^{\mathrm{II}}/\mathrm{d}t, \end{aligned} \tag{2.48}$$

The species-specific term in this equation is

$$\begin{aligned} -\mathrm{d}M_X/\mathrm{d}t &= -(\mu_X - \mu_X^*)\, \mathrm{d}n_X^{\mathrm{II}}/\mathrm{d}t \\ &= RT\left(t_X^+ - t_X^-\right) \ln\left(t_X^+ / t_X^-\right) = D_X \end{aligned} \tag{2.49}$$

and $D_X \geq 0$ for all p_X.

The relation to the stochastic theory does not generalize here for cases without detailed balance except for the approach to equilibrium, [1].

Acknowledgement. This chapter is based on parts of ref. [1], 'Thermodynamics far from Equilibrium: Reaction with Multiple Stationary States.'

References

1. J. Ross, K.L.C. Hunt, P.M. Hunt, J. Chem. Phys. **88**, 2719–2729 (1988)
2. P. Glansdorff, I. Prigogine, *Thermodynamic Theory of Structure, Stability, and Fluctuations* (Wiley, New York, 1971)
3. D. Jou, J. Casas-Vázquez, G. Lebon, *Extended Irreversible Thermodynamics*, 3rd ed., (Springer, Berlin Heidelberg New York, 2001)
4. B.C. Eu, *Kinetic Theory and Irreversible Thermodynamics* (Wiley, New York, 1992)
5. N.G. Van Kampen, *Stochastic Processes in Physics and Chemistry* (North-Holland, Amsterdam, 1981)
6. P. Glansdorff, G. Nicolis, I. Prigogine, Proc. Natl. Acad. Sci. USA **71**, 197–199 (1974)

3

Thermodynamic State Function for Single and Multivariable Systems

3.1 Introduction

In Chap. 2 we obtained a thermodynamic state function ϕ^*, (2.13), valid for single variable non-linear systems, and (2.6), valid for single variable linear systems. We shall extend the approach used there to multi-variable systems in Chap. 4 and use the results later for comparison with experiments on relative stability. However, the generalization of the results in Chap. 2 for multi-variable linear and non-linear systems, based on the use of deterministic kinetic equations, does not yield a thermodynamic state function. In order to obtain a thermodynamic state function for multi-variable systems we need to consider fluctuations, and now turn to this analysis [1].

We start with the master equation [2]

$$\frac{\partial}{\partial t} P_{\mathbf{X}}(\mathbf{X},t) = \sum_{\mathbf{r}} \left[W(\mathbf{X}-\mathbf{r},\mathbf{r}) P_{\mathbf{X}}(\mathbf{X}-\mathbf{r},t) - W(\mathbf{X},\mathbf{r}) P_{\mathbf{X}}(\mathbf{X},t) \right], \quad (3.1)$$

in which $P_{\mathbf{X}}$ is the probability distribution of finding $\mathbf{X}$ particles (molecules) in a given volume, and $W(\mathbf{X},\ \mathbf{r})$ is the transition probability due to reaction from $\mathbf{X}$ to $\mathbf{X}+\mathbf{r}$ particles. Now we do a Taylor expansion of the term $W(\mathbf{X}-\mathbf{r},\ \mathbf{r})\, P(\mathbf{X}-\mathbf{r},\mathrm{t})$ around $\mathbf{X}$

$$\begin{aligned} W(\mathbf{X}-\mathbf{r},\mathbf{r}) P_{\mathbf{X}}(\mathbf{X}-\mathbf{r},t) &= W(\mathbf{X},\mathbf{r}) P_{\mathbf{X}}(\mathbf{X},t) \\ &+ \sum_{m=1}^{\infty} \frac{(-1)^m}{m!} \left(\mathbf{r}\cdot\nabla_{\mathbf{X}}\right)^m \left[W(\mathbf{X},\mathbf{r}) P_{\mathbf{X}}(\mathbf{X},t)\right], \end{aligned} \quad (3.2)$$

and introduce the concentration vector:

$$\mathbf{x} = \mathbf{X}/V; \quad (3.3)$$

then we have the reduced relations

$$\nabla_{\mathbf{X}} = \frac{1}{V}\nabla_{\mathbf{x}},\ P_{\mathbf{X}}(\mathbf{X},t)\,\Delta\mathbf{X} = P_{\mathbf{x}}(\mathbf{x},t)\,\mathrm{d}\mathbf{x},\ \Delta\mathbf{X}=1,\ w(\mathbf{x},\mathbf{r};V) = \frac{1}{V} W(\mathbf{x}V,\mathbf{r}), \quad (3.4)$$

where V is the volume of the system. We substitute these relations into (3.2) and obtain

$$W(\mathbf{X}-\mathbf{r},\mathbf{r})\,P_{\mathbf{X}}(\mathbf{X}-\mathbf{r},t) = \mathrm{d}\mathbf{x}W(\mathbf{x}V,\mathbf{r})\,P_{\mathbf{x}}(\mathbf{x},t) + \mathrm{d}\mathbf{x}\sum_{m=1}^{\infty}\frac{(-1)^m}{m!}\left(\frac{1}{V}\mathbf{r}\cdot\nabla_{\mathbf{x}}\right)^m [W(\mathbf{x}V,\mathbf{r})\,P_{\mathbf{x}}(\mathbf{x},t)]\,. \tag{3.5}$$

Next we introduce the momentum operator

$$\hat{\mathbf{p}} = -\frac{1}{V}\nabla_{\mathbf{x}} \tag{3.6}$$

with which we can write

$$1+\sum_{m=1}^{\infty}\frac{(-1)^m}{m!}\left(\frac{1}{V}\mathbf{r}\cdot\nabla_{\mathbf{x}}\right)^m \ldots\ldots = 1+\sum_{m=1}^{\infty}\frac{1}{m!}(\mathbf{r}\cdot\hat{\mathbf{p}})^m \ldots\ldots = \exp(\mathbf{r}\cdot\hat{\mathbf{p}})\ldots\ldots \tag{3.7}$$

The master equation becomes:

$$\frac{1}{V}\frac{\partial}{\partial t}P_{\mathbf{x}}(\mathbf{x},t) = \sum_{\mathbf{r}} w(\mathbf{x},\mathbf{r};V)\,[\exp(\mathbf{r}\cdot\hat{\mathbf{p}})-1]\,P_{\mathbf{x}}(\mathbf{x},t) = \hat{H}_{+}(\mathbf{x},\hat{\mathbf{p}})\,P_{\mathbf{x}}(\mathbf{x},t)\,, \tag{3.8}$$

where we have defined the Hamiltonian operator (2–4)

$$\hat{H}_{+}(\mathbf{x},\hat{\mathbf{p}})\ldots = \sum_{\mathbf{r}} w(\mathbf{x},\mathbf{r};V)\,[\exp(\mathbf{r}\cdot\hat{\mathbf{p}})\ldots\ldots-1]\,. \tag{3.9}$$

Thus we have formally, and exactly, converted the master equation to a Schroedinger equation. This has the substantial advantage that we can apply well-known approximations in quantum mechanics to obtain solutions to the master equation. In particular we refer to the W.K.B. approximation valid for semiclassical cases, those for which Planck's constant formally approaches zero. The equivalent limit for (3.8) is that of large volumes (large numbers of particles). Hence we seek a stationary solution of (3.8), that is the time derivative of $P_{\mathbf{X}}(\boldsymbol{X},t)$ is set to zero, of the form

$$P_s^{(n)}(\mathbf{X}) = C^{(n)}\exp[-VS_n(\mathbf{X})] \tag{3.10}$$

where S_n will be shown to be the classical action of a fluctuational trajectory accessible from the nth stable stationary state. We substitute (3.10) into the stationary part of (3.8) and obtain

$$H\left[\mathbf{x},\frac{\partial S_n(\mathbf{x})}{\partial \mathbf{x}}\right] = 0, \tag{3.11}$$

the equation satisfied by $S_n(\mathbf{X})$ with the Hamiltonian function (not operator)

$$H(\mathbf{x},\mathbf{p}) = \sum_{\mathbf{r}} w(\mathbf{x},\mathbf{r})[\exp(\mathbf{r}\cdot\mathbf{p}) - 1], \tag{3.12}$$

and the boundary condition

$$S_n(\mathbf{x}_n^{\mathrm{s}}) = 0. \tag{3.13}$$

These equations show that it is the classical action S_n that satisfies the Hamiltonian–Jacoby equation (3.11) with coordinate $\mathbf{x}$, momentum $\mathbf{p} = \partial S_n(\mathbf{x})/\partial\mathbf{x}$, and Hamiltonian equal to zero (stationary condition). The Hamiltonian equations of motion for the system are

$$\frac{\mathrm{d}\mathbf{x}}{\mathrm{d}t} = \sum_{\mathbf{r}} \mathbf{r}w(\mathbf{x},\mathbf{r})\exp(\mathbf{r}.\mathbf{p}) \tag{3.14}$$

and

$$\frac{\mathrm{d}\mathbf{p}}{\mathrm{d}t} = -\sum_{\mathbf{r}}[\exp(\mathbf{r}.\mathbf{p}) - 1]\frac{\partial w(\mathbf{x},\mathbf{r})}{\partial\mathbf{x}}. \tag{3.15}$$

From these relations we determine the action

$$S_n(\mathbf{x}) = \int_{-\infty}^{0} \mathrm{d}t\ \mathbf{p}\cdot\mathrm{d}\mathbf{x}/\mathrm{d}t \tag{3.16}$$

for the fluctuational trajectory starting at the nth stable stationary state $\mathbf{x}_n^{\mathrm{s}}$ with $\mathbf{p} = 0$ at $t = -\infty$ and ending at $\mathbf{x}$ at $t = 0$.

3.2 Linear Multi-Variable Systems

Let us apply these equations to a linear reaction system [1]

$$A \underset{k_2}{\overset{k^1}{\rightleftharpoons}} X \underset{k_4}{\overset{k^3}{\rightleftharpoons}} Y \underset{k_6}{\overset{k^5}{\rightleftharpoons}} B \tag{3.17}$$

run in an apparatus as in Fig. 1.1 of Chap. 1, with the pressures of A and B held constant. The deterministic kinetic equations are

$$\left.\frac{\mathrm{d}X}{\mathrm{d}t}\right|_{\mathrm{det}} = k_1A - (k_2 + k_3)X + k_4Y = -[(k_2 + k_3)(X - X_{\mathrm{s}})] - k_4(Y - Y_{\mathrm{s}})] \tag{3.18}$$

and

$$\left.\frac{\mathrm{d}Y}{\mathrm{d}t}\right|_{\mathrm{det}} = k_3X - (k_4 + k_5)Y + k_6B = -[(k_4 + k_5)(Y - Y_{\mathrm{s}})] - k_3(X - X_{\mathrm{s}})]. \tag{3.19}$$

Table 3.1. Mechanics steps and **r** values for $A \to X \to Y \to B$

Step	X	Y	$W(\mathbf{r}, \mathbf{X})$
1	+1	0	$k_1 A$
2	−1	0	$k_2 X$
3	−1	+1	$k_3 X$
4	+1	−1	$k_4 Y$
5	0	−1	$k_5 Y$
6	0	+1	$k_1 B$

For the reaction mechanism in (3.17) there are six elementary reaction steps with different values of **r** and transition probabilities, and these are listed in Table 3.1, taken from [1].

Now we use the Hamiltonian equations of motion to obtain the fluctuational trajectories:

$$\begin{aligned}
\left.\frac{\mathrm{d}X}{\mathrm{d}t}\right|_{\mathrm{fl}} &= k_1 A - \exp(p_x) - k_2 X \exp(-p_x) - k_3 X \exp(p_y - p_x) \\
&\quad + k_4 Y \exp(p_x - p_y), \\
\left.\frac{\mathrm{d}Y}{\mathrm{d}t}\right|_{\mathrm{fl}} &= k_3 X - \exp(p_y - p_x) - k_4 Y \exp(p_x - p_y) \\
&\quad - k_5 Y \exp(-p_y) + k_6 B \exp(p_y), \\
\frac{\mathrm{d}p_x}{\mathrm{d}t} &= k_2[1 - \exp(-p_x)] + k_3[1 - \exp(p_y - p_x)], \\
\frac{\mathrm{d}p_y}{\mathrm{d}t} &= k_4[1 - \exp(p_x - p_y)] + k_5[1 - \exp(-p_y)].
\end{aligned} \tag{3.20}$$

We see that we obtain coupled non-linear equations for this reaction mechanism with linear rate laws. There are several ways of solving these equations. We show that Hamilton's equations have the solution

$$p_x = \ln(X/X_{\mathrm{s}}) \tag{3.21}$$

and

$$p_y = \ln(Y/Y_{\mathrm{s}}), \tag{3.22}$$

which gives the momenta on the fluctuational trajectory in terms of the stable stationary state concentrations of the deterministic kinetic equations. To show this we substitute (3.21) and (3.22) into the first two equations of (3.20) and obtain

$$\left.\frac{\mathrm{d}X}{\mathrm{d}t}\right|_{\mathrm{fl}} = (k_2 + k_3)(X - X_{\mathrm{s}}) - k_3[X_{\mathrm{s}}(Y - Y_{\mathrm{s}})/Y_{\mathrm{s}}]$$

and

$$\left.\frac{\mathrm{d}Y}{\mathrm{d}t}\right|_{\mathrm{fl}} = (k_4 + k_5)(Y - Y_{\mathrm{s}}) - k_4[Y_{\mathrm{s}}(X - X_{\mathrm{s}})/X_{\mathrm{s}}]. \tag{3.23}$$

Next we substitute (3.21) and (3.22) into the last two equations of (3.20) with the result

$$\begin{aligned}\frac{\mathrm{d}p_x}{\mathrm{d}t} &= -(1/X)[k_2(X - X_\mathrm{s}) - k_3(X_\mathrm{s}Y/Y_\mathrm{s} - X)] \\ &= (1/X)\frac{\mathrm{d}X}{\mathrm{d}t}\Big|_{\mathrm{fl}}, \\ \frac{\mathrm{d}p_y}{\mathrm{d}t} &= -(1/Y)[k_4(Y_\mathrm{s}X/X_\mathrm{s} - Y) + k_5(Y - Y_\mathrm{s})] \\ &= (1/Y)\frac{\mathrm{d}Y}{\mathrm{d}t}\Big|_{\mathrm{fl}},\end{aligned} \tag{3.24}$$

which agrees with differentiation of (3.21) and (3.22) with respect to time. The Hamiltonian of the system, (3.12), is explicitly

$$\begin{aligned}H(\mathbf{x},\mathbf{p}) = V^{-1}\{&k_1A[\exp(p_x) - 1] + k_2X[\exp(-p_x) - 1] \\ &+k_3X[\exp(p_y - p_x)] + k_4Y[\exp(p_x - p_y) - 1] \\ &+k_5Y[\exp(-p_y) - 1] + k_6B[\exp(p_y) - 1]\}.\end{aligned} \tag{3.25}$$

Substituting (3.21) and (3.22) into (3.25) yields

$$\begin{aligned}H(\mathbf{x},\mathbf{p}) &= V^{-1}\{k_1A(X/X_\mathrm{s} - 1) + k_2X(X_\mathrm{s}/X - 1) \\ &\quad +k_3X[X_\mathrm{s}Y/(XY_\mathrm{s}) - 1] + k_4Y[XY_\mathrm{s}/(X_\mathrm{s}Y) - 1] \\ &\quad +k_5Y(Y_\mathrm{s}/Y - 1) + k_6B(Y/Y_\mathrm{s} - 1)\}, \\ &= V^{-1}[(k_1A - k_2X_\mathrm{s} - k_3X_\mathrm{s} + k_4Y_\mathrm{s})(X/X_\mathrm{s}) \\ &\quad +(k_3X_\mathrm{s} - k_4Y_\mathrm{s} - k_5Y_\mathrm{s} + k_6B)(Y/Y_\mathrm{s}) \\ &\quad -(k_1A - k_2X_\mathrm{s} - k_5Y_\mathrm{s} + k_6B)].\end{aligned} \tag{3.26}$$

In the second equation of (3.26) the terms have been arranged into three groups according to

$$\mathrm{d}X_\mathrm{s}/\mathrm{d}t = 0, \quad \mathrm{d}Y_\mathrm{s}/\mathrm{d}t = 0, \tag{3.27}$$

and

$$\mathrm{d}X_\mathrm{s}/\mathrm{d}t + \mathrm{d}Y_\mathrm{s}/\mathrm{d}t = 0, \tag{3.28}$$

all of which vanish and hence the Hamiltonian vanishes, $H(\mathbf{x},\mathbf{p}) = 0$. Equation (3.23) determine the fluctuational trajectory in the space of concentrations (X, Y). This trajectory is in general not the same as the time-reversed deterministic path from given initial values of (X, Y) to the stable stationary state, except for the case for which the concentrations (A, B) have their equilibrium ratio. The master equation for this linear system does not have detailed balance unless (A, B) have their equilibrium ratio. For a discussion of detailed balance, microscopic reversibility and mesoscopic balance see the end of Chap. 18.

From the above relations we find the action, (3.16), given by

$$S_n(\mathbf{x}) = \int_{-\infty}^{0} \mathrm{d}t \left[\ln(X/X_\mathrm{s}) \frac{\mathrm{d}x}{\mathrm{d}t} |_\mathrm{fl} + \ln(Y/Y_\mathrm{s}) \frac{\mathrm{d}y}{\mathrm{d}t} |_\mathrm{fl} \right]$$
$$= \int_\mathrm{s}^{x,y} [\ln(X/X_\mathrm{s})\mathrm{d}x + \ln(Y/Y_\mathrm{s})\mathrm{d}y]. \tag{3.29}$$

From the second of these equations we see that the integrand is an exact differential and hence the action is independent of the path of integration in concentration space. The action is a state function. This result has been reported in a number of publications [2, 5–7].

The physical interpretation of the action in (3.29) comes from consideration of the free energy M, see Chap. 1, (2.21) for the three compartments, Fig. 1.1 in Chap. 1

$$M = G^\mathrm{I} + A^\mathrm{II} + G^\mathrm{III}. \tag{3.30}$$

For differential changes in A, X, Y, B the differential change in M is

$$\mathrm{d}M = \mu_A \ \mathrm{d}n_A + \mu_X \ \mathrm{d}n_X + \mu_Y \ \mathrm{d}n_Y + \mu_B \ \mathrm{d}n_B. \tag{3.31}$$

The differential excess free energy change $\mathrm{d}\phi$ is the difference between $\mathrm{d}M$ the system with arbitrary concentrations of X and Y and $\mathrm{d}M_\mathrm{s}$ for the system in the stationary state. Hence we have

$$\mathrm{d}\phi = (\mu_X - \mu_X^\mathrm{s})\mathrm{d}n_X + (\mu_Y - \mu_Y^\mathrm{s})\mathrm{d}n_Y. \tag{3.32}$$

When we compare (3.32) with the second equation of (3.29) we see that

$$\mathrm{d}\phi/kT = V \ \mathrm{d}S, \tag{3.33}$$

This important physical result was first given in [1]: the mathematical concept of the action can be identified with the thermodynamic excess work.

On a fluctuational trajectory the differential excess free energy is positive and zero at a stable stationary state. We show this by considering the differential action

$$\left.\frac{\mathrm{d}S}{\mathrm{d}t}\right|_\mathrm{fl} = \mathbf{p} \cdot \mathrm{d}\mathbf{x}/\mathrm{d}t|_\mathrm{fl} - H(\mathbf{x}, \mathbf{p}) = \sum_\mathbf{r} w(\mathbf{x}, \mathbf{r})[(\mathbf{r} \cdot \mathbf{p}) \exp(\mathbf{r} \cdot \mathbf{p}) - \exp(\mathbf{r} \cdot \mathbf{p}) + 1]. \tag{3.34}$$

The transition probabilities $\omega(\mathbf{x}, \mathbf{r})$ are all positive and the square bracket is larger than zero except for $\mathbf{p} = 0$, that is at the stable stationary state. Therefore we have

$$\frac{\mathrm{d}S}{\mathrm{d}t} |_\mathrm{fl} \geq 0 \tag{3.35}$$

and hence from (3.36) the excess differential free energy $\mathrm{d}\phi$ is positive in general and zero at stationary states.

Suppose we prepare this system at a given (x, y) and let it proceed along the deterministic trajectory back to the stationary state. Along this path $\mathrm{d}\phi$ is negative which follows from the deterministic variation of the action in time

$$\frac{\mathrm{d}S}{\mathrm{d}t}|_{\mathrm{det}} = \nabla S \cdot \frac{\mathrm{d}\mathbf{x}}{\mathrm{d}t}|_{\mathrm{det}} = \mathbf{p} \cdot \frac{\mathrm{d}\mathbf{x}}{\mathrm{d}t}|_{\mathrm{det}} = \mathbf{p} \cdot \frac{\mathrm{d}\mathbf{x}}{\mathrm{d}t}|_{\mathrm{det}} - H(\mathbf{x}, \mathbf{p})$$

$$= \sum_{\mathbf{r}} w(\mathbf{x}, \mathbf{r})[(\mathbf{r} \cdot \mathbf{p}) - \exp(\mathbf{r} \cdot \mathbf{p}) + 1], \tag{3.36}$$

which holds since the Hamiltonian is zero. For all real values of $\mathbf{r} \cdot \mathbf{p}$ the square bracket in the second line of (3.36) is negative unless $\mathbf{p} = 0$. And therefore

$$\frac{\mathrm{d}S}{\mathrm{d}t}|_{\mathrm{det}} \leq 0 \tag{3.37}$$

with the equality holding only at a stationary state. Hence an excess work is required to move a system from a stable stationary state, and excess work can be done by a system relaxing towards a stable stationary state. The action and the excess work are both Liapunov functions; they serve for non-equilibrium systems the role the Gibbs free energy serves for systems going to equilibrium.

We note here that (3.35) and (3.37) hold for non-linear multi-variable systems as well; no assumption of a linear reaction mechanism was made in their derivation.

For linear systems in (3.29) and (3.33) the first derivatives of the excess work with respect to species numbers or concentrations x, y are zero at each stationary state, and the second derivative is equal to or greater than zero at each stable stationary state, and equal to or less than zero at each unstable stationary state, in exact parallel for single variable systems, (2.17)–(2.19).

The fluctuational trajectory away from a stationary state to a given point in concentration space (x, y) may differ from the deterministic path from that point back to the stationary state, for systems without detailed balance. Of course, the free energy change must vanish for a closed loop in the space of (A,B,X,Y) but need not vanish for a closed loop in the restricted space of (x, y).

3.3 Nonlinear Multi-Variable Systems

We turn next to consideration of a non-linear multi-variable system, for example the model

$$A + (m-1)X \underset{k_2}{\overset{k_1}{\rightleftharpoons}} mX,$$

$$rX + (s-1)Y \underset{k_4}{\overset{k_3}{\rightleftharpoons}} (r-1)X + sY,$$

$$nY \underset{k_6}{\overset{k_5}{\rightleftharpoons}} (n-1)Y + B. \tag{3.38}$$

The stationary distribution is given by (3.10) and (3.16) with $\mathbf{p}$ and $\mathrm{d}\mathbf{x}/\mathrm{d}t$ obtained from solutions of Hamilton's equations. We now choose our reference state not as in Chap. 2, but in analogy with 3.21 and 3.22 we identify a reference state by using the equations

$$\begin{aligned} p_x &= \ln(X/X^0) \\ p_y &= \ln(Y/Y^0). \end{aligned} \tag{3.39}$$

Equation (3.38) yield unique values of (X^0, Y^0) in the absence of certain crossings of fluctuational trajectories in the (X, Y) space, called 'caustics', see [8]. There may be more than one fluctuational trajectory which starts at $\mathbf{p} = 0$ at a stable stationary state and passes through a given (X, Y). These trajectories will have different values of $\mathbf{p}$ and the one with the lowest value of $\mathbf{p}$ will determine the action in the thermodynamic limit, the contributions from other trajectories vanishing in that limit.

Hence we find for the action the expressions

$$\begin{aligned} S_n(\mathbf{x}) &= \int_{-\infty}^{0} \mathrm{d}t \left[\ln(X/X^0) \frac{\mathrm{d}x}{\mathrm{d}t} |_{\mathrm{fl}} + \ln(Y/Y^0) \frac{\mathrm{d}y}{\mathrm{d}t} |_{\mathrm{fl}} \right] \\ &= \int_{\mathrm{s}}^{x,y} [\ln(X/X^0)\mathrm{d}x + \ln(Y/Y^0)\mathrm{d}y] \\ &= 1/RT \int_{\mathrm{s}}^{x,y} [(\mu_X - \mu_X{}^0)\mathrm{d}x + (\mu_Y - \mu_{Y^0})\mathrm{d}y] \\ &= (1/VkT) \int_{\mathrm{s}}^{X,Y} \mathrm{d}\phi^0 . \end{aligned} \tag{3.40}$$

Here the reference state (X^0, Y^0) replaces the starred reference state of Chap. 2 (see (2.11)). The important point is that the action and the excess work in (3.40) are state functions for single and multi-variable systems. Both X^0 and Y^0 are functions of X and Y in general, but the integrand in (3.40) is an exact differential, because $\mathbf{p}$ is the gradient of the action, (3.16). For the starred reference state the excess work is a state function only for single variable systems.

The fluctuational trajectory away from a stationary state to a given point in concentration space (X, Y) in general differs from the deterministic path from that point back to the stationary state for systems without detailed balance. We show this in some calculations for the Selkov model; in (3.38) we take $m = n = r = 1, s = 3$; other parameters are given in [1], p. 4555. Figure 3.1 gives some results of these calculations.

S1 and S3 are stable stationary states (stable foci); S2 denotes an unstable stationary state. The solid line from S2 to S3 indicates the deterministic trajectory. The other solid line through S2 is the deterministic separatrix, that is the line that separates deterministic trajectories, on one side going towards S2 and on the other side going towards S3. The dotted lines are fluctuational

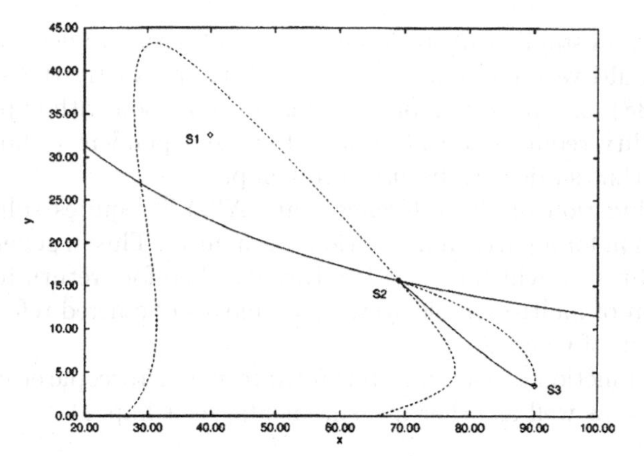

Fig. 3.1. From [1] S1 and S3 are stable stationary states (stable foci); S2 denotes an unstable stationary state. The solid line from S2 to S3 indicates the deterministic trajectory. The other solid line through S2 is the deterministic separatrix, that is the line that separates deterministic trajectories, on one side going towards S2 and on the other side going towards S3. The dotted lines are fluctuational trajectories: one from S3 to S2 and the others proceeding from S2 in two different directions. The fluctuational trajectory need not differ so much from the reverse of the deterministic trajectory, as we shall show for some sets of parameters in Chap. 4.

trajectories: one from S3 to S2 and the others proceeding from S2 in two different directions. The fluctuational trajectory need not differ so much from the reverse of the deterministic trajectory, as we shall show for some sets of parameters in Chap. 4.

For one-variable systems the fluctuational trajectory away from the stationary state is the same as the deterministic trajectory back to the stationary state. Therefore for such systems ϕ^* equals ϕ^0.

In summary, we define the state function ϕ^0 with the use of (3.40)

$$\phi^0 = \int_{\mathrm{s}}^{X,Y} d\phi^0 \tag{3.41}$$

and list the following results (compare with the results listed in Chap. 2 for single variables systems).

ϕ^0 is a state function. It is a potential for the stationary probability distribution of the master equation, and is a Lyapunov function in the domain of each stable stationary state. See also [8–14]. It is an extremum at stationary states; a miminum (zero) at stable stationary states, a maximum at unstable stationary states, (3.35). For a fluctuational trajectory ϕ^0 increases away from the stable stationary state, (3.35); for a deterministic trajectory towards a stable stationary state it decreases, (3.36). The first derivative of ϕ^0 is larger than zero at each stable stationary state, smaller than zero at each unstable stationary state. The function ϕ^0 provides necessary and sufficient criteria for the existence and stability of stationary states. ϕ^0 serves to determine relative stability of multi-variable homogeneous systems in exactly the same way as shown in (2.38) for single variable systems. Comparison with experiments on relative stability requires consideration of space-dependent (inhomogeneous) systems and that subject is discussed in Chap. 5.

The specification of the reference state X^0, Y^0 requires solution of the master equation for a particular reaction mechanism. This in general demands numerical solutions, which can be lengthy. We therefore return in Chap. 4 to a presentation of multi-variable systems by means of starred reference states, in continuation of Chap. 2.

The state function ϕ^0 can be determined from macroscopic electrochemical measurements, as well as other measurements, see Chap. 11.

Acknowledgement. This chapter is based largely on [1].

References

1. B. Peng, K.L.C. Hunt, P.M. Hunt, A. Suárez, J. Ross, J. Chem. Phys. **102**, 4548–4562 (1995)
2. I. Oppenheim, K.E. Shuler, G.H. Weiss, *Stochastic Processes in Chemical Physics: The Master Equation* (MIT, Cambridge, MA, 1977). C.W. Gardiner, *Handbook of Stochastic Methods* (Springer, Berlin Heidelberg New York, 1990.) N.G. van Kampen, *Stochastic Processes in Physics and Chemistry* (North-Holland, New York, 1992)
3. R. Kubo, K. Matsuo, K. Kitahara, J. Stat. Phys. **9**, 51–96 (1973)
4. K. Kitahara, Adv. Chem. Phys. **29**, 85–111 (1973)
5. M.I. Dykman, E. Mori, J. Ross, P.M. Hunt, J. Chem. Phys. **100**, 5735–5750 (1994)
6. P.M. Hunt, K.L.C. Hunt, J. Ross, J. Chem. Phys. **92**, 2572–2581 (1990)
7. G. Hu, Phys. Rev. A. **36**, 5782–5790 (1987)
8. G. Nicolis, A. Babloyantz, J. Chem. Phys. **51**, 2632–2637 (1969)
9. R. Graham, T. Tél, Phys. Rev. A. **33**, 1322–1337 (1986)
10. R.S. Maier, D.L. Stein, Phys. Rev. Lett. **69**, 3691–3695 (1992)
11. H.R. Jauslin, Physica A. **144**, 179–191 (1987)
12. H.R. Jauslin, J. Stat. Phys. **42**, 573–585 (1986)
13. M.I. Freidlin, A.D. Wentzell, *Random Perturbations of Dynamical Systems* (Springer, Berlin Heidelberg New York, 1984)
14. M.G. Crandall, L.C. Evans, P.L. Lions, Trans. AMS. **282**, 487–502 (1984)

4

Continuation of Deterministic Approach for Multivariable Systems

In Chap. 2 we analyzed single variable linear and non-linear systems with single and multiple stable stationary states by use of the deterministic equations of chemical kinetics. We introduced species-specific affinities and the concept of an excess work; with these we showed the existence of a thermodynamic state function ϕ^* and compiled its many interesting properties, see (2.15–2.19), including its relation to fluctuations as given by the stationary solution of the master equation, (2.34). We continue this approach here by turning to systems with more than one intermediate, [1].

We begin again with linear reactions and consider the reaction mechanism

$$\boldsymbol{A} \underset{k_2}{\overset{k_1}{\rightleftarrows}} \boldsymbol{X} \underset{k_4}{\overset{k_3}{\rightleftarrows}} \boldsymbol{Y} \underset{k_6}{\overset{k_5}{\rightleftarrows}} \boldsymbol{B}, \tag{4.1}$$

run in the apparatus, Fig. 1.1 of Chap. 1. Both X and Y are present in volume II; the pressures of A and B are held constant. The hybrid free energy M of the system (see (2.21)) is

$$M = G_{\mathrm{I}} + A_{\mathrm{II}} + G_{\mathrm{III}}, \tag{4.2}$$

The pressures of X and Y vary in time according to the mass-action kinetic equations

$$\begin{aligned} \frac{\mathrm{d}p_X}{\mathrm{d}t} &= t_X^+ - t_X^-, \\ \frac{\mathrm{d}p_Y}{\mathrm{d}t} &= t_Y^+ - t_Y^-, \end{aligned} \tag{4.3}$$

where the reaction rates are

$$\begin{aligned} t_X^+ &= k_1 p_A + k_4 p_Y, \\ t_Y^+ &= k_3 p_X + k_6 p_B, \\ t_X^- &= (k_2 + k_3) p_X, \\ t_Y^- &= (k_4 + k_5)\, p_Y. \end{aligned} \tag{4.4}$$

There is a unique stationary state for this system in which the pressures of X and Y are

$$\begin{aligned} p_X^s &= (k_1 p_A + k_4 p_Y^s) \,/\, (k_2 + k_3) \\ p_Y^s &= (k_3 p_X^s + k_6 p_B) \,/\, (k_4 + k_5) \end{aligned} \tag{4.5}$$

We identify species-specific affinities for X and Y, $(\mu_X^s - \mu_X)$ and $(\mu_Y^s - \mu_Y)$, both of which vanish at the stationary state. They are related to the reaction rates by the equations

$$\begin{aligned} (\mu_X - \mu_X^s) &= -RT \ln \left(t_X^{+s} / t_X^- \right) \\ (\mu_Y - \mu_Y^s) &= -RT \ln \left(t_Y^{+s} / t_Y^- \right) . \end{aligned} \tag{4.6}$$

We can find an interpretation of the species-specific affinities with a thought experiment carried out in the apparatus shown in Fig. 4.1.

Each reservoir is separated from the central reaction chamber by a membrane permeable only to the species in that reservoir. C, C′, C″ are catalysts for the three reactions in (4.1), all of which occur only in the central chamber. From [1].

Consider a change of the thermodynamic state of the entire system in which one mole of X at the pressure p_X is formed from Δn_A moles of A, Δn_B

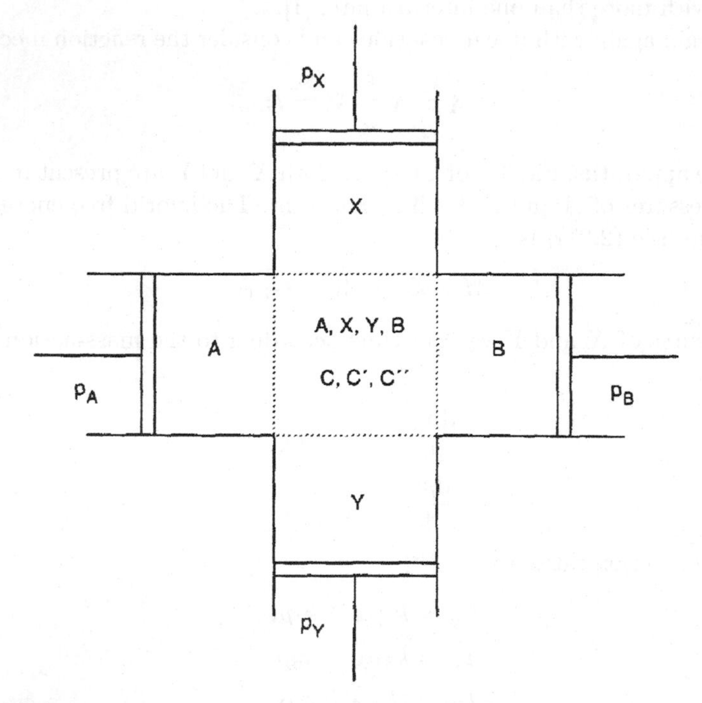

Fig. 4.1. Schematic drawing of a four piston model. The pressures of A, X, Y, B are held constant by external forces on the respective pistons

moles of B, and Δn_Y moles of Y. The change in the free energy of the entire system is

$$\Delta M\left(p_X\right) = -\mu_A \Delta n_A - \mu_B \Delta n_B - \mu_Y \Delta n_Y + \mu_X. \tag{4.7}$$

For the four piston apparatus the change in M, (4.7), is the same as the change in G, since all reservoirs are at constant pressure and temperature. Similarly we may consider the change in M at the pressure of the stationary state, for which we have

$$\Delta M\left(p_X^{\mathrm{s}}\right) = -\mu_A \Delta n_A - \mu_B \Delta n_B - \mu_Y \Delta n_Y + \mu_X^{\mathrm{s}}. \tag{4.8}$$

Hence we see that the species-specific affinity is an excess free energy, an excess work,

$$\Delta M\left(p_X\right) - \Delta M\left(p_X^{\mathrm{s}}\right) = \left(\mu_X - \mu_X^{\mathrm{s}}\right) \tag{4.9}$$

with the relations to the kinetic expressions given by (4.6). A similar interpretation holds for $(\mu_Y - \mu_Y{}^{\mathrm{s}})$.

The function

$$\phi = \int_{n_X^{\mathrm{s}},\, n_Y^{\mathrm{s}}}^{n_X,\, n_Y} \left(\mu_X - \mu_X^{\mathrm{s}}\right) \mathrm{d}n_X + \left(\mu_Y - \mu_Y^{\mathrm{s}}\right) \mathrm{d}n_Y \tag{4.10}$$

is a thermodynamic state function and a Liapunov function for the system. The stability property listed in Chap. 2, (2.17) and (2.18), hold for the linear two variable system and the function in (4.10) determines the fluctuations in the stationary state of the master equation for this system, the analogue of (2.34). Next we study a non-linear system with multiple intermediates and multiple stationary

$$\begin{aligned} &A + (m-1)\, X \underset{k_2}{\overset{k_1}{\rightleftarrows}} mX, \\ &qX + (r-1)\, Y \underset{k_4}{\overset{k_3}{\rightleftarrows}} (q-1)\, X + rY, \\ &nY \underset{k_6}{\overset{k_5}{\rightleftarrows}} (n-1)\, Y + B. \end{aligned} \tag{4.11}$$

states, which, for example for $m, n, q = 1$, $r = 3$, is the Selkov model which may have multiple stationary states and limit cycles. The reactions occur in an apparatus like Fig. 4.2, in volume II.

The macroscopic, the deterministic, kinetic equations for this system are

$$\begin{aligned} \mathrm{d}p_X/\mathrm{d}t &= t_X^+ - t_X^-, \\ t_X^+ &= k_1 p_A p_X^{m-1} + k_4 p_X^{q-1} p_Y^r, \\ t_X^- &= k_2 p_X^m + k_3 p_X^q p_Y^{r-1}, \\ \mathrm{d}p_Y/\mathrm{d}t &= t_Y^+ - t_Y^-, \\ t_Y^+ &= k_3 p_X^q p_Y^{r-1} + k_6 p_Y^{n-1} p_B, \\ t_Y^- &= k_5 p_Y^n + k_4 p_X^{q-1} p_Y^r \end{aligned} \tag{4.12}$$

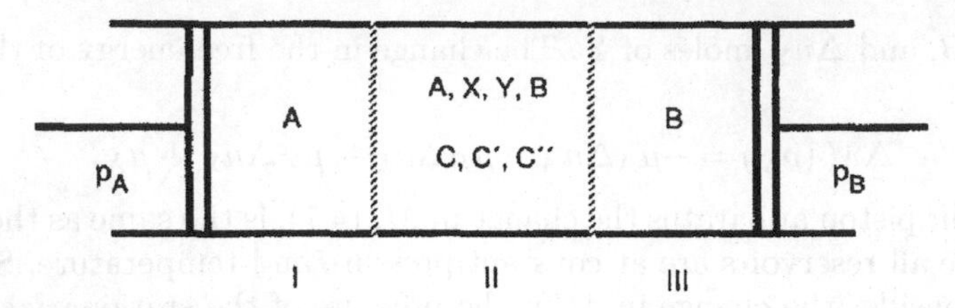

Fig. 4.2. Schematic diagram for two-piston model. From [1]

for reactions of ideal gases (and for ideal liquids if expressed in terms of concentrations).

We need again the concept of thermodynamic and kinetic indistinguishability of the given non-linear system with a specified linear system: the state variables, equilibrium constants, and quantities derivable from them are identical and so are the kinetic terms. The species-specific affinities, marked by a star, must satisfy two conditions: $(\mu_X - \mu_X^*)$ and $(\mu_Y - \mu_Y^*)$ must each vanish at all stationary states, and the values of each bracket must be identical in all instantaneously indistinguishable systems. This can be achieved with the substitutions

$$\begin{aligned} k_1 p_X^{m-1} &= k_1^\dagger, \\ k_2 p_X^{m-1} &= k_2^\dagger, \\ k_3 p_X^{q-1} p_Y^{r-1} &= k_3^\dagger, \\ k_4 p_X^{q-1} p_Y^{r-1} &= k_4^\dagger, \\ k_5 p_Y^{n-1} &= k_5^\dagger, \\ k_6 p_Y^{n-1} &= k_6^\dagger. \end{aligned} \tag{4.13}$$

Here the dagger denotes the instantaneously indistinguishable linear system. Since $p_X = p_X^\dagger$, we have

$$\begin{aligned} p_X^* &= p_X^{s\dagger}, \\ p_Y^* &= p_Y^{s\dagger}. \end{aligned} \tag{4.14}$$

The substitution of (4.14) into (4.5) yields the equations necessary for obtaining the two unknowns p_X^* and p_Y^*

$$\begin{aligned} p_X^* &= \frac{\left[k_1 (p_X)^{m-1} p_A + k_4 (p_X)^{q-1} (p_Y)^{r-1} p_Y^*\right]}{\left[k_2 (p_X)^{m-1} + k_3 (p_X)^{q-1} (p_Y)^{r-1}\right]}, \\ p_Y^* &= \frac{\left[k_3 (p_X)^{q-1} (p_Y)^{r-1} p_X^* + k_6 (p_Y)^{n-1} p_B\right]}{\left[k_4 (p_X)^{q-1} (p_Y)^{r-1} + k_5 (p_Y)^{n-1}\right]}. \end{aligned} \tag{4.15}$$

It is useful to define the quantities

$$t_X^{+*} = k_1 (p_X)^{m-1} p_A + k_4 (p_X)^{q-1} (p_Y)^{r-1} p_Y^*,$$
$$t_Y^{+*} = k_3 (p_X)^{q-1} (p_Y)^{r-1} p_X^* + k_6 (p_Y)^{n-1} p_B \tag{4.16}$$

and we recall (2.12) which we can use here.

Hence as in Chap. 2, we may define a function ϕ

$$\phi = \int_{n_X^s\; n_Y^s}^{n_X,\; n_Y} (\mu_X - \mu_X^*)\, dn_X + (\mu_Y - \mu_Y^*)\, dn_Y$$
$$= -V^{\mathrm{II}} \int_{p_X^s, p_Y^s}^{P_X, p_Y} \ln \left(t_X^{+*}/t_X^-\right) dp_X + \ln \left(t_Y^{+*}/t_Y^-\right) dp_Y. \tag{4.17}$$

If p_A and p_B are chosen such that their ratio differs from the equilibrium constant of the system and either X or Y are in an autocatalytic step then the starred quantities are functions of X, Y and ϕ is not a state function. Neither is ϕ in (4.17) a solution to the stationary form of the master equation, although in some cases it can be a useful approximation (see below). This differs from the earlier results where we found ϕ to be a state function for any system for which equilibrium is the only stationary state; for any non-autocatalytic system for which there is only one stationary state; and for any system with only one intermediate, whether there is autocatalysis or not. In all these three case ϕ provides a solution to the time independent master equation.

As is the cases in earlier chapters, the function ϕ in (4.17) is zero at stationary states, increases on removal from stable stationary states and decreases from any initial given state on its approach to the nearest stable stationary state along a deterministic kinetic trajectory. These specifications make ϕ a Liapunov function in the vicinity of stable stationary states, which indicates the direction of the deterministic motion. Hence for every variation from a stable stationary state we have

$$(\delta\phi)_{p_{X^s}, p_{Y^s}} > 0. \tag{4.18}$$

This equation serves as a necessary and sufficient criterion for the existence and stability of stationary states for non-autocatalytic and auto-catalytic stationary states in multi-variable systems.

The hypothesis that the function ϕ, (4.17), provides a solution to the stationary master equation, time-independent, requires a guess: we know that the deterministic trajectory is the most probable path from some initial X, Y to the closest stable stationary state; what, however, is the most probable fluctuational trajectory from that stable state to X, Y? The guess is that this fluctuational trajectory is the reverse of the deterministic relaxation from X, Y to the stationary state. This guess (approximation) is sometimes good, but can also be quite bad. An example of each is given in the next two figures.

In Fig. 4.3, [2], we plot a cut through the stationary solution of the master equation for selected parameters of the Selkov model vs. the variable Y; the dotted line is a numerical solution of the probability distribution and the solid line is that distribution calculated from (4.15–4.18) with the approximation described in this paragraph and the same parameters for the Selkov model. The approximation gives a reasonable estimate. A different impression is gathered from the plot shown in Fig. 4.4: A most probable fluctuational trajectory obtained from numerical integration of the stationary solution of the master

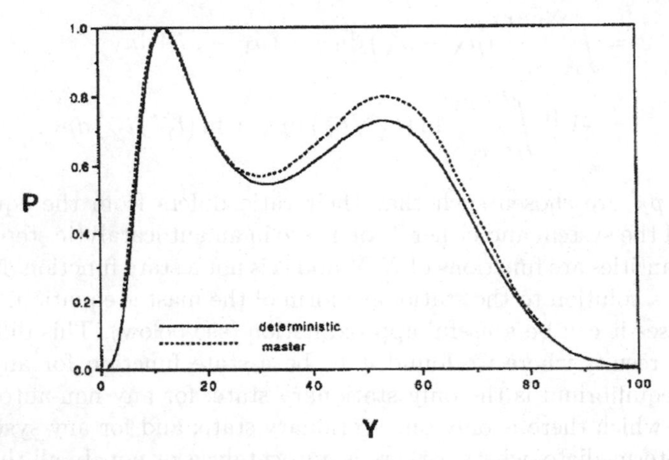

Fig. 4.3. Plot of the stationary probability distribution of the Selkov model: a cut at constant X vs. Y. The parameters used are given in [2], see the caption to Fig. 8 in that reference. From [2]

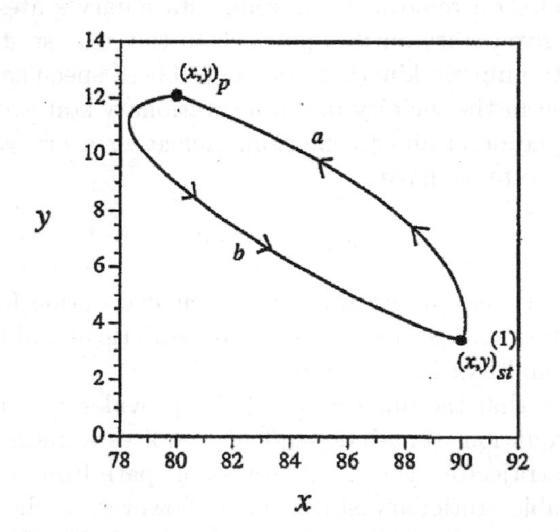

Fig. 4.4. Plot in the concentration space of the variables x, y of the Selkov model: (a) optimal fluctuational trajectory from a stable stationary state $(x, y)_{\mathrm{st}}$ to a given point $(x, y)_{\mathrm{p}}$ and (b) the deterministic return to the stationary state. From [3]

equation, again for the Selkov model, with parameters given in [3], p. 5745. There is a clear difference between that most probable fluctuational trajectory away from the stable stationary state and the deterministic return to that state.

We shall use the results of this chapter when we come to the analysis of reaction diffusion systems and a discussion of the relative stability of two stationary stable states based on theory and experiments.

Acknowledgement. This chapter is based on the results reported in [1–3].

References

1. J. Ross, K.L.C. Hunt, P.M. Hunt, J. Chem. Phys. **96**, 618–629 (1992)
2. Q. Zheng, J. Ross, K.L.C. Hunt, P.M. Hunt, J. Chem. Phys. **96**, 630–640 (1992)
3. M.I. Dykman, E. Mori, J. Ross, P.M. Hunt, J. Chem. Phys. **100**, 5735–5750 (1994)

5

Thermodynamic and Stochastic Theory of Reaction–Diffusion Systems

Relative Stability of Multiple Stationary States

So far we have considered only homogeneous reaction systems in which concentrations are functions of time only. Now we turn to inhomogeneous reaction systems in which concentrations are functions of time and space. There may be concentration gradients in space and therefore diffusion will occur. We shall formulate a thermodynamic and stochastic theory for such systems [1]: first we analyze one-variable systems and then two- and multi-variable systems, with two or more stable stationary states, and then apply the theory to study relative stability of such multiple stable stationary states. The thermodynamic and stochastic theory of diffusion and other transport processes is given in Chap. 8.

Let us take a one-variable system, such as the Schlögl model

$$A + 2X \leftrightarrows 3X, \quad X \leftrightarrows B \tag{5.1}$$

for which the deterministic kinetics in a homogeneous system is

$$\frac{\mathrm{d}X}{\mathrm{d}t} = t^+(X) - t^-(X), \tag{5.2}$$

where X denotes the total number of molecules of that species in a given constant volume and t^+, t^- are the kinetic fluxes that increase and decrease X, respectively.

For inhomogeneous systems the number densities (concentrations) are functions of time and spatial coordinates. In a one-dimensional system with spatial coordinate z we discretize the space into many boxes, labelled with $\ldots i-1,\ i,\ i+1$, see Fig. 5.1.

The increase and decrease of the number of particles X_i are due to reaction and diffusion into and out of box i and can be written

$$\frac{\mathrm{d}X_i}{\mathrm{d}t} = t^+(X_i) - t^-(X_i) + t^+_{\mathrm{D}i} - t^-_{\mathrm{D}i}, \tag{5.3}$$

with the first two terms on the rhs denoting increase and decrease due to reaction, and the third and fourth term are fluxes of diffusion into and out of box i

$$t_{\mathrm{D}_i}^{+} = d(X_{i-1} + X_{i+1})$$
$$t_{\mathrm{D}_i}^{-} = 2dX_i, \tag{5.4}$$

where d is a constant diffusion coefficient. In the continuous limit we have

$$\frac{\mathrm{d}x(z)}{\mathrm{d}t} = t^{+}[x(z)] - t^{-}[x(z)] + D\frac{\partial^2 x(z)}{\partial z^2}, \tag{5.5}$$

where $D = l^2 d$ is the diffusivity of the system and l is the length of a box. Changes in the reactant A and product B take place at constant temperature and pressure (p_A, p_B) and changes in X at constant temperature and volume; hence the differential hybrid free energy $\mathrm{d}M_i$ for box i and the auxiliary reservoirs is

$$\mathrm{d}M_i = \mu(A_i)\mathrm{d}A_i + \mu(X_i)\mathrm{d}X_i + \mu(B_i)\mathrm{d}B_i, \tag{5.6}$$

where A_i and B_i are the number of molecules of reactant A and product B in box i.

Next, we seek a linear system which is thermodynamically and kinetically equivalent to the system described by (5.3)

$$\frac{\mathrm{d}X_i^{\mathrm{L}}}{\mathrm{d}t} = a_i - b_i X_i^{\mathrm{L}} + t_{\mathrm{D}i}^{+\mathrm{L}} - t_{\mathrm{D}i}^{-\mathrm{L}}. \tag{5.7}$$

The superscript L denotes the Linear equivalent system. The instantaneous equivalencies are guaranteed by the requirements

$$a_i = t^{+}(X_i) \quad \text{and} \quad b_i = \frac{t^{-}(X_i)}{X_i}. \tag{5.8}$$

The stationary solution of (5.7) is

$$X_i^{\mathrm{LS}} = X_i^{\mathrm{L}} \frac{a_i + t_{\mathrm{D}i}^{+\mathrm{LS}}}{b_i X_i^{\mathrm{L}} + t_{\mathrm{D}i}^{-\mathrm{L}}}, \tag{5.9}$$

where

$$t_{\mathrm{D}i}^{+\mathrm{LS}} = d\left(X_{i-1}^{\mathrm{LS}} + X_{i+1}^{\mathrm{LS}}\right). \tag{5.10}$$

The difference between the free energy change $\mathrm{d}M$ at an arbitrary state X_i^{L} and that at the stationary state X_i^{LS} under the conditions of given A_i, B_i, $\mathrm{d}A_i$, $\mathrm{d}B_i$, and $\mathrm{d}X_i$ is the excess work for box i in this equivalent linear system

$$\begin{aligned} \tilde{d}\Phi_i\left(X_i^{\mathrm{L}}\right) &= \mathrm{d}M_i - \mathrm{d}M_i^{\mathrm{s}} \\ &= \left\{\mu\left(X_i^{\mathrm{L}}\right) - \mu\left(X_i^{\mathrm{LS}}\right)\right\}\mathrm{d}X_i^{\mathrm{L}} = kT\ln\frac{X_i^{\mathrm{L}}}{X_i^{\mathrm{LS}}}\,\mathrm{d}X_i^{\mathrm{L}}, \end{aligned} \tag{5.11}$$

where

$$\left\{\mu\left(X_i^{\mathrm{L}}\right) - \mu\left(X_i^{\mathrm{LS}}\right)\right\} \tag{5.12}$$

is the driving force towards the stationary state. On the lhs of (5.11) the curl on d indicates that the differential is inexact.

In a non-linear system the driving force is the potential difference between state X_i and a reference state X_i^*, which is the stationary state of the equivalent linear system at the specified value of X_i. Thus from (5.8) and (5.9) we have

$$X_i^* = X_i^{\mathrm{LS}} = X_i \frac{t^+(X_i) + t_{\mathrm{D}i}^{+*}}{t^-(X_i) + t_{\mathrm{D}i}^-}, \tag{5.13}$$

where

$$t_{\mathrm{D}i}^{+*} = \mathrm{d}(X_{i-1}^* + X_{i+1}^*).$$

The excess work for box i of the non-linear system is

$$\tilde{\mathrm{d}}\Phi_i(X_i) = \{\mu(X_i) - \mu(X_i^*)\}\mathrm{d}X_i = kT \ln \frac{X_i}{X_i^*}\mathrm{d}X_i, \tag{5.14}$$

which is obtained by substitution of superscript 'S' in (5.11) by a superscript star '*'. The total excess work is the sum of that work in all the boxes

$$\begin{aligned} \Phi[\{X_i\}] &= kT \sum_i \int^{x_i} \ln \frac{X_i'}{X_i'^*}\mathrm{d}X_i' \\ &= kt \sum_i \int^{x_i} \ln \frac{t^-(X_i') + t_{\mathrm{D}i}^-}{t^+(X_i') + t_{\mathrm{d}i}^{+*}}\mathrm{d}X_i'. \end{aligned} \tag{5.15}$$

The number of X molecules in each box i is an independent variable X_i and hence the present reaction–diffusion system is isomorphic to a multivariable homogeneous system. To evaluate Φ in (5.15) a path of integration needs to be specified because Φ is not a state function.

There are two limiting cases which we can easily check. In the homogeneous limit there is no diffusion; hence we have only one box for which

$$t_D^+ = t_D^- = 0; \tag{5.16}$$

and therefore (5.15) reduces to

$$\Phi(X) = kT \int^x \ln \frac{X'}{X'^*}\mathrm{d}X' = kT \int^x \ln \frac{t^-(X')}{t^+(X')}\mathrm{d}X'. \tag{5.17}$$

This is the result for a homogeneous chemical reaction, see (2.12) and (2.13). In the limit of no reaction we set

$$t^+(X_i) = t^-(X_i) = 0, \tag{5.18}$$

and (5.15) reduces to

$$\begin{aligned} \Phi[\{X_i\}] &= kT \sum_i \int^{x_i} \ln \frac{X_i'}{X_i'^*} \mathrm{d}X_i' \\ &= kT \sum_i \int^{x_i} \ln \frac{t_{\mathrm{D}i}^-}{t_{\mathrm{D}i}^{+*}} \mathrm{d}X_i'; \end{aligned} \tag{5.19}$$

this agrees with the analysis for diffusion only, see appendix.

In the limit of a continuous distribution in the spatial variable z in (5.15) that equation becomes

$$\Phi[x(z)] = S \int \mathrm{d}z \left[\int^{x(z)} \mathrm{d}x' \{\mu(x') - \mu(x'^*)\} \right], \tag{5.20}$$

where S is the area of the system perpendicular to the z-axis.

Reaction–diffusion systems, linear or not, can be mapped into multi-variable reaction systems, as stated after (5.15). For such multi-variable reaction systems which can be linearized in the vicinity of a stable stationary state, we have at that state

$$\left. \frac{\mathrm{d}\Phi}{\mathrm{d}X_i} \right|_{\mathrm{det}}^{\mathrm{s}} = 0, \tag{5.21}$$

where the subscript 'det' denotes the deterministic path as the path of integration. The time derivative of Φ satisfies the equation

$$\frac{\mathrm{d}}{\mathrm{d}t} \Phi \leq 0, \tag{5.22}$$

which is proven in Appendix at the end of this chapter. Φ is a minimum at stable stationary states and is a Liapunov function in the vicinity of such states.

For linear reaction mechanisms Φ can be shown to be the solution of the stationary master equation (see Appendix A in [1]); we shall have no need for it.

5.1 Reaction–Diffusion Systems with Two Intermediates

We now consider reaction-diffusion systems with two intermediates and multiple stationary states, which may be nodes or foci.[1] For a real eigenvalue that approach is monotonic; for a complex eigenvalue with negative real part that approach is one of damped oscillations. In the absence of cross diffusion the deterministic rate equations in one dimension, z, are

[1] A node (focus) is a stationary state which is approached in time described by an eigenvalue which is real (complex, with negative real part).

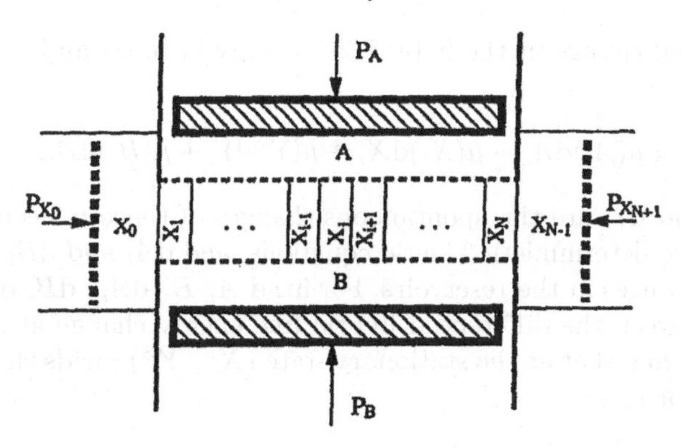

Fig. 5.1. Schematic apparatus for reaction–diffusion system in one spatial dimension. The boxes 1 through N are separated from a constant-pressure reservoir of A by a membrane permeable only to A, and similarly for the reservoir of B. From [1]

$$\begin{aligned}\frac{\mathrm{d}x}{\mathrm{d}t} &= t_x^+(x,y) - t_x^-(x,y) + D_x\frac{\partial^2 x}{\partial z^2};\\ \frac{\mathrm{d}y}{\mathrm{d}t} &= t_y^+(x,y) - t_y^-(x,y) + D_y\frac{\partial^2 y}{\partial z^2}.\end{aligned} \tag{5.23}$$

with the usual notation. With discretization of the system into boxes, as in Fig. 5.1, we write

$$\begin{aligned}\frac{\mathrm{d}X_i}{\mathrm{d}t} &= t_x^+(X_i,Y_i) - t_x^-(X_i,Y_i) + d_x(X_{i+1} + X_{i-1} - 2X_i);\\ \frac{\mathrm{d}Y_i}{\mathrm{d}t} &= t_y^+(X_i,Y_i) - t_y^-(X_i,Y_i) + d_y(Y_{i+1} + Y_{i-1} - 2Y_i).\end{aligned} \tag{5.24}$$

5.1.1 Linear Reaction Systems

For linear reaction systems, for example

$$A \underset{k_2}{\overset{k_1}{\rightleftarrows}} X \underset{k_4}{\overset{k_3}{\rightleftarrows}} Y \underset{k_6}{\overset{k_5}{\rightleftarrows}} B, \tag{5.25}$$

the chemical fluxes must take the form

$$\begin{aligned}&t_X^+(X_i,Y_i) = a_X + b_X Y_i;\ t_X^-(X_i,Y_i) = c_X X_i;\\ &t_Y^+(X_i,Y_i) = a_y + b_y X_i;\ t_Y^-(X_i,Y_i) = c_y Y_i\end{aligned} \tag{5.26}$$

For the example in (5.25) we have

$$\begin{aligned}&t_X^+(X_i,Y_i) = k_1 A + k_4 Y_i,\ t_X^-(X_i,Y_i) = (k_2 + k_3)X_i,\\ &t_Y^+(X_i,Y_i) = k_6 B + k_3 X_i \text{ and } t_Y^-(X_i,Y_i) = (k_4 + k_5)Y_i\end{aligned} \tag{5.27}$$

The differential change in the hybrid free energy in box i and and the reservoirs is

$$\mathrm{d}M_i = \mu(A_i)\mathrm{d}A_i + \mu(X_i)\mathrm{d}X_i + \mu(Y_i)\mathrm{d}Y_i + \mu(B_i)\mathrm{d}B_i, \tag{5.28}$$

where $\mathrm{d}X_i$ and $\mathrm{d}Y_i$ are the spontaneous changes of those molecules in box i as given by the deterministic kinetic equations, and $\mathrm{d}A_i$ and $\mathrm{d}B_i$ are the corresponding changes in the reservoirs. For fixed A, B, $\mathrm{d}A_i$, $\mathrm{d}B_i$ $\mathrm{d}X_i$, $\mathrm{d}Y_i$ the difference between the differential hybrid free energy change at an arbitrary state $(X_i,\ Y_i)$ and that at the stationary state $\left(X_i^{\mathrm{S}},\ Y_i^{\mathrm{S}}\right)$ yields the differential excess work for box i

$$\begin{aligned}\mathrm{d}\Phi_i &= \mathrm{d}M_i - \mathrm{d}M_i^{\mathrm{S}} \\ &= \{\mu(X_i) - \mu\left(X_i^{\mathrm{S}}\right)\}\mathrm{d}X_i + \{\mu(Y_i) - \mu\left(Y_i^{\mathrm{S}}\right)\}\mathrm{d}Y_i \\ &= kT\ln\frac{X_i}{X_{i\mathrm{S}}}\mathrm{d}X_i + kT\ln\frac{Y_i}{Y_{i\mathrm{S}}}\mathrm{d}Y_i,\end{aligned} \tag{5.29}$$

where

$$\left(X_i^{\mathrm{S}}, Y_i^{\mathrm{S}}\right) = \left[\frac{a_x + b_x Y_i^{\mathrm{S}} + d_x(X_{i+1}^{\mathrm{S}} + X_{i-1}^{\mathrm{S}})}{(c_x + 2d_x)}, \frac{a_y + b_y X_i^{\mathrm{S}} + d_y\left(Y_{i+1}^{\mathrm{S}} + Y_{i-1}^{\mathrm{S}}\right)}{(c_y + 2d_y)}\right] \tag{5.30}$$

is the stationary solution of the linear reaction diffusion system. We can then write $\mathrm{d}\Phi_i$ as

$$\begin{aligned}d\Phi_i = {}& kT\ln\frac{(c_x + 2d_x)X_i}{a_x + b_x Y_i^{\mathrm{S}} + d_x(X_{i+1}^{\mathrm{S}} + X_{i-1}^{\mathrm{S}})}\mathrm{d}X_i \\ &+ kT\ln\frac{(c_y + 2d_y)Y_i}{a_y + b_y X_i^{\mathrm{S}} + d_y(Y_{i+1}^{\mathrm{S}} + Y_{i-1}^{\mathrm{S}})}\mathrm{d}Y_i.\end{aligned} \tag{5.31}$$

The total excess work is the sum over all boxes i

$$\begin{aligned}\Phi[\{X_i, Y_i\}] = \sum_i \Bigg(& \int^{X_i} \mathrm{d}X_i'\{\mu(X_i') - \mu(X_i'^{\mathrm{S}})\} \\ & + \int^{Y_i} \mathrm{d}Y_i'\{\mu(Y_i') - \mu\left(Y_i'^{\mathrm{S}}\right)\}\Bigg).\end{aligned} \tag{5.32}$$

For linear systems the integral in (5.31) is path independent and therefore Φ is a state function. The first derivative of Φ with respect to X_i is

$$\frac{\partial}{\partial X_i}\Phi[\{X_i, Y_i\}] = \mu(X_i) - \mu(X_i^{\mathrm{S}}), \tag{5.33}$$

which is zero at the stable stationary state; and similarly the first derivative with respect to Y_i. The second derivative of Φ with respect to X_i is

$$\frac{\partial^2}{\partial X_i^2}\Phi[\{X_i, Y_i\}] = RT\frac{1}{X_i} > 0 \tag{5.34}$$

that is larger than zero, and hence Φ is a minimum at the stable stationary state. The derivative with respect to time is

$$\frac{\mathrm{d}}{\mathrm{d}t}\Phi[\{X_i, Y_i\}] = \sum_i \left(\frac{\mathrm{d}X_i}{\mathrm{d}t}\{\mu(X_i) - \mu(X_i^{\mathrm{S}})\} + \frac{\mathrm{d}Y_i}{\mathrm{d}t}\{\mu(Y_i) - \mu(Y_i^{\mathrm{S}})\}\right). \tag{5.35}$$

The right hand side of (5.34) is negative semidefinite, so that the system tends towards the minimum of Φ, that is towards the stable stationary state. Thus the function Φ is a Liapunov function of the system. Further, Φ satisfies the stationary solution of the master equation in the thermodynamic limit. All these properties assure that the function Φ provides nessecary and sufficient conditions for the existence and stability of stationary states.

5.1.2 Non-Linear Reaction Mechanisms

Next we analyse chemical reaction systems with autocatalytic steps in which the kinetic reaction terms may be non-linear functions of the concentrations of the intermediate species. We have in mind, once again, a reaction mechanism as shown in (4.11). We require that the number of X molecules changes by ± 1 or 0 in each elementary reaction step, and similarly for Y. For each set of (X_i, Y_i) we can construct at each instance a thermodynamically and kinetically equivalent system; the mapping from the non-linear to the equivalent linear system is unique. The linear equivalent system is chosen as shown in (5.26), but now the coefficients satisfy the relations given in Table 5.1.

Table 5.1. Relations of the terms in the rate equations of a non-linear system, (4.11) and (4.12), to the kinetically equivalent linear system for each set of variables, see (5.26). From (1)

Linear		Non-linear
a_{xi}	=	$k_1 A X_i^{m-1}$
b_{xi}	=	$k_4 X_i^{q-1} Y_i^{r-1}$
c_{xi}	=	$k_2 X_i^{m-1} + k_3 X_i^{q-1} Y_i^{r-1}$
a_{yi}	=	$k_6 B Y_i^{n-1}$
b_{yi}	=	$k_3 X_i^{q-1} Y_i^{r-1}$
c_{yi}	=	$k_5 Y_i^{n-1} + k_4 X_i^{q-1} Y_i^{r-1}$

The reference state $(X_i^*,\ Y_i^*)$ is defined by the equations

$$\begin{aligned} X_i^* &= \frac{a_{xi} + b_{xi}Y_i^* + d_x(X_{i+1}^* + X_{i-1}^*)}{(c_{xi} + 2d_x)} \\ &= \frac{k_1 A X_i^{m-1} + k_4 X_i^{q-1} Y_i^{r-1} Y_i^* + d_x(X_{i+1}^* + X_{i-1}^*)}{k_2 X_i^{m-1} + k_3 X_i^{q-1} Y_i^{r-1} + 2d_x} \end{aligned} \tag{5.36}$$

$$\begin{aligned} Y_i^* &= \frac{a_{yi} + b_{yi}X_i^* + d_y(Y_{i+1}^* + Y_{i-1}^*)}{(c_{yi} + 2d_y)} \\ &= \frac{k_6 B Y_i^{n-1} + k_3 X_i^{q-1} Y_i^{r-1} X_i^* + d_y(Y_{i+1}^* + Y_{i-1}^*)}{k_5 Y_i^{n-1} + k_4 X_i^{q-1} Y_i^{r-1} + 2d_y}. \end{aligned} \tag{5.37}$$

As the system approaches a stationary state the starred variables approach their values of the stationary state. For the Selkov model the stationary state of the linear equivalent system is

$$\begin{aligned} (X_i^{\prime S}, Y_i^{\prime S}) &= (X_i^*, Y_i^*) \\ &= \left(\frac{k_1 A + k_4 Y_i^2 Y_i^* + d_x(X_{i+1}^* + X_{i-1}^*)}{(k_2 + k_3 Y_i^2 + 2d_x)}, \right. \\ &\qquad \left. \frac{k_6 B + k_3 Y_i^2 X_i^* + d_y(Y_{i+1}^* + Y_{i-1}^*)}{(k_5 + k_4 Y_i^2 + 2d_y)} \right), \end{aligned} \tag{5.38}$$

where the prime indicates the corresponding value of the linear equivalent system.

The differential excess work for the equivalent linear system is given in (5.29), and the instantaneous differential excess work for the non-linear system is

$$\begin{aligned} \tilde{\mathrm{d}}\Phi_i &= \{\mu(X_i) - \mu(X_i^*)\}\mathrm{d}X_i + \{\mu(Y_i) - \mu(Y_i^*)\}\mathrm{d}Y_i \\ &= kT \ln \frac{X_i}{X_i^*}\mathrm{d}X_i + kT \ln \frac{Y_i}{Y_i^*}\mathrm{d}Y_i, \end{aligned} \tag{5.39}$$

where again the curl on d on the lhs indicates an inexact differential. The quantities in brackets on the first line of (5.39) are the species-specific driving forces for species X and Y towards the reference state $(\boldsymbol{X}_i^*, \boldsymbol{Y}_i^*)$. Formally the function Φ for non-linear systems can be determined by changing the superscript 'S', which indicates the stationary state for the linear system, (5.29) to a '*' defined in (5.36) and (5.37) with corresponding relations given in Table 5.1. The total excess work is

$$
\begin{aligned}
\Phi &= \sum_i \int^{X_i Y_i} \tilde{\mathrm{d}}\Phi_i \\
&= \sum_i \int^{X_i Y_i} \{\mu(X_i) - \mu(X_i^*)\}\mathrm{d}X_i \\
&\quad + \{\mu(Y_i) - \mu(Y_i^*)\}\mathrm{d}Y_i \\
&= kT \sum_i \int^{X_i Y_i} \ln \frac{X_i}{X_i^*}\mathrm{d}X_i + \ln \frac{Y_i}{Y_i^*}\mathrm{d}Y_i .
\end{aligned} \tag{5.40}
$$

The function Φ is determined when we choose a path of integration, such as the deterministic path or its reverse. Φ is zero at the stationary state and its first derivatives at the stationary state are zero

$$
\left.\frac{\partial \Phi}{\partial X_i}\right|^{\mathrm{S}} = \left.\frac{\partial \Phi}{\partial Y_i}\right|^{\mathrm{S}} = 0. \tag{5.41}
$$

Since the non-linear system is indistinguishable from the instantaneously equivalent linear system, we have

$$
\left.\frac{\mathrm{d}\Phi}{\mathrm{d}t}\right|_{\text{non-linear}} = \left.\frac{\mathrm{d}\Phi}{\mathrm{d}t}\right|_{\text{linear}} . \tag{5.42}
$$

The rhs of this equation is negative semi-definite (Appendix) and we have the result

$$
\left.\frac{\mathrm{d}\Phi}{\mathrm{d}t}\right|_{\text{non-linear}} \leq 0 \tag{5.43}
$$

at every state $(X_i,\ Y_i)$; the equality holds only at the stationary state. Φ decreases in time and at the stationary state it is a minimum. Hence Φ is a Liapunov function and serves to determine the necessary and sufficient conditions for the existence and stability of stationary states for the systems under consideration here.

5.1.3 Relative Stability of Two Stable Stationary States of a Reaction–Diffusion System

Consider a reaction system with two stable stationary states. We wish to consider the issue of relative stability of these two states for stated external constraints. We do that routinely for systems at or near equilibrium, say water (liquid) and water (vapour), at one atmosphere pressure and temperature T. For $T < 100°\mathrm{C}$ the water (l) is more stable and the Gibbs free energy of water (l) is less than that of water vapour (v); if two phases of water, one liquid, the other vapour, are placed in contact with each other then the vapour will condense to form liquid, the more stable phase. If $T > 100°\mathrm{C}$ then the reverse is true, and $G_{\mathrm{v}} < G_{\mathrm{l}}$. A similar argument holds for two stable stationary states of a reaction diffusion system.

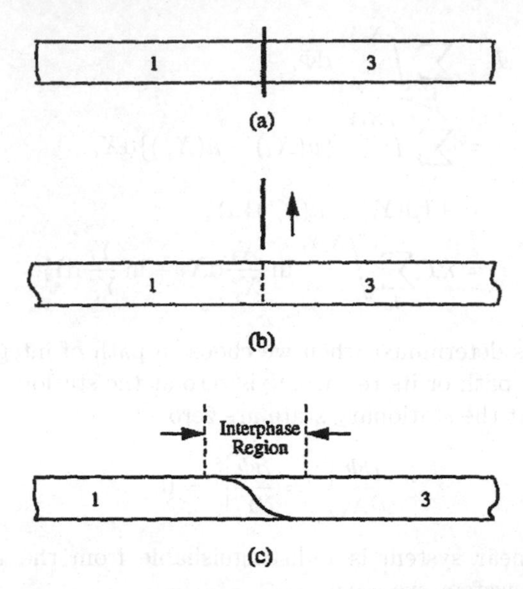

Fig. 5.2. Schematic apparatus for determining relative stability of two stable stationary states of a reaction–diffusion system. For description see text. From [1]

Imagine the following schematic apparatus. Take a semi-infinite tube and fill it with one stable stationary state, call it 1, and take another semi-infinite tube and fill it with the other stable stationary state, call it 3. Both tubes are at the same external constraints, of temperature, pressure and concentration of species. Place the tubes lengthwise together, see Fig. 5.2a, at first with a partition between them.

Then remove the partition, Fig. 5.2b; reaction and diffusion will occur and during some transient time a reaction diffusion front may form in an interphase region, Fig. 5.2c, and travel into the less stable state. At external constraints corresponding to equistablity the velocity of propagation of the reaction–diffusion front is zero.

We now show that the concept of excess work as developed in this and earlier chapters serves as a criterion of equistability, in the same way as the Gibbs free energy serves that purpose for systems at equilibrium. Once a reaction diffusion front is established, see Fig. 5.3 we may calculate the excess work of creating that front from phase 1 and, separately calculate the excess work of establishing that front from phase 3.

If these two excess works are equal than we expect equistability and zero velocity of front propagation. If the excess work to form the front from phase 1 is less than that necessary to form the front from phase 3, then we expect phase one to be more stable than phase 3. To do that calculation we divide the interphase region into N boxes of a specified length L, with the boundary conditions of the concentrations on the left side of the interphase

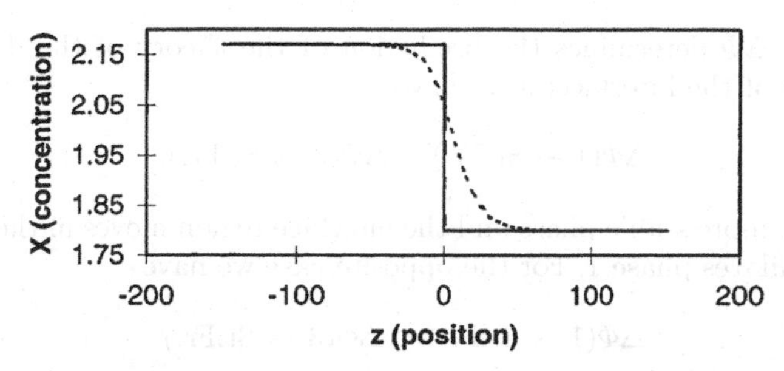

Fig. 5.3. Plot of concentration vs. position z. The initial concentration profile is shown by the *solid line*; the space with negative z is filled initially with stationary state 1, the space with positive z is filled initially with stationary state 3. The *dotted line* indicates the interface region. From [1]

region set at the concentration of state 1, and those on the right side with the concentrations of state 3. The initial concentrations in the interphase region are set to those of phase 1 3. In each case diffusion and reaction will occur in each box i, and each box follows a different deterministic path from the initial stationary state 1 or 3 to the stable front condition in box i. We obtain these paths from integrating numerically the $2N$ ordinary differential equation

$$\begin{aligned}\frac{\mathrm{d}X_i}{\mathrm{d}t} &= t_x^+(X_i, Y_i) - t_x^-(X_i, Y_i) + d_x(X_{i+1} + X_{i-1} - 2X_i),\\ \frac{\mathrm{d}Y_i}{\mathrm{d}t} &= t_y^+(X_i, Y_i) - t_y^-(X_i, Y_i) + d_y(Y_{i+1} + Y_{i-1} - 2Y_i),\\ &(i = 1, \ldots, N).\end{aligned} \tag{5.44}$$

Along the calculated trajectories we evaluate $(X_i^*,\ Y_i^*)$ from (5.35) and (5.36). Then we can obtain the excess work from the last line of (5.40)

$$\Delta\Phi(1 \to 3) = kT \sum_i \int_{\text{phase 1}}^{\text{phase 3}} \ln \frac{X_i}{X_i^*}\mathrm{d}X_i + \ln \frac{Y_i}{Y_i^*}\mathrm{d}Y_i. \tag{5.45}$$

We can split the expression on the rhs into two parts

$$\begin{aligned}\Delta\Phi(1 \to 3) = \;& kT \sum_i \int_{\text{phase 1}}^{\text{St.Fr.}} \left\{ \ln \frac{X_i}{X_i^*}\mathrm{d}X_i + \ln \frac{Y_i}{Y_i^*}\mathrm{d}Y_i \right\}\\ & - kT \sum_i \int_{\text{phase 3}}^{\text{St.Fr.}} \left\{ \ln \frac{X_i}{X_i^*}\mathrm{d}X_i + \ln \frac{Y_i}{Y_i^*}\mathrm{d}Y_i \right\}\\ =\;& \Delta\Phi(1 \to \text{St.Fr.}) - \Delta\Phi(3 \to \text{St.Fr.}).\end{aligned} \tag{5.46}$$

The sign of $\Delta\Phi$ determines the prediction of the theory of the direction of propagation of the interface: if we have

$$\Delta\Phi(1 \to \text{St.Fr.}) > \Delta\Phi(3 \to \text{St.Fr.}), \tag{5.47}$$

then 3 is the more stable phase and the interface region moves in the direction which annihilates phase 1. For the opposite case we have

$$\Delta\Phi(1 \to \text{St.Fr.}) < \Delta\Phi(3 \to \text{St.Fr.}) \tag{5.48}$$

for which phase 1 is more stable and the interface region moves in the direction which annihilates phase 3. For the case

$$\Delta\Phi(1 \to \text{St.Fr.}) = \Delta\Phi(3 \to \text{St.Fr.}) \tag{5.49}$$

the thermodynamic theory predicts equistability of the two phases and the interface does not move.

5.1.4 Calculation of Relative Stability in a Two-Variable Example, the Selkov Model

In this section we compare the predictions of the thermodynamic theory for relative stability in a two-variable example, the Selkov model, with the results obtained from numerical integration of the reaction diffusion equation. The model, constructed for early studies of glycolysis, has two variables, X and Y, and two constant concentrations. The reaction mechanism is

$$\begin{aligned} A &\leftrightarrow X \\ X + 2Y &\leftrightarrow 3Y \\ Y &\leftrightarrow B \end{aligned} \tag{5.50}$$

and the reaction–diffusion equations in one dimension (z) are

$$\begin{aligned} \frac{\partial x}{\partial t} &= k_1 A + k_4 y^3 - k_2 x - k_3 y^2 x + D_x \frac{\partial^2 x}{\partial z^2}, \\ \frac{\partial y}{\partial t} &= k_6 B + k_3 y^2 x - k_5 y - k_4 y^3 + D_y \frac{\partial^2 y}{\partial z^2}, \end{aligned} \tag{5.51}$$

where k_1 is the rate coefficient of the first reaction in (5.49) in the forward direction, k_2 for the first reaction in the backward reaction, and so on; x and y are the concentrations of X and Y, respectively. For certain ranges of the parameter $k_6 B$ and the other parameters (see [1] p. 3451) there are three stationary states, two of which are stable and labelled 1 and 3, and one of which is unstable and labelled 2. In an arrangement as in Fig. 5.2, initially the left side is in phase 1, the right side in phase 3. The direction of propagation is determined by all the kinetic parameters, the concentrations of

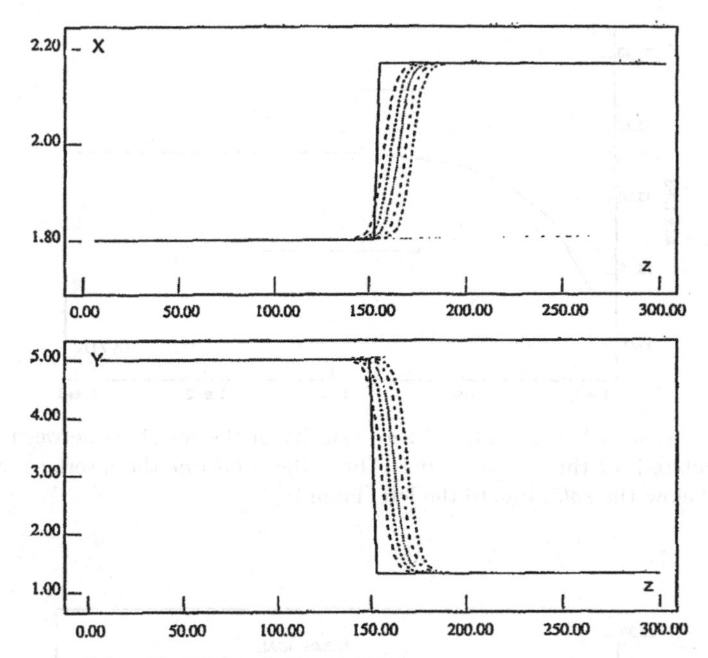

Fig. 5.4. Plots of concentration profiles of X and Y vs. distance z during the front propagation to the right in the Selkov model. The *solid line* is the initial concentration profile; the *dotted lines* are concentration profiles with a spacing of 500 in arbitrary time units. For values of parameters see the caption to Fig. 4 in [1]. From [1]

A and B, and the ratio of the diffusivities $\delta = D_y/D_x$. In Fig. 5.4 we show an example of the solutions of the reaction diffusion equations for this system.

For certain values of the parameters a stationary front is obtained; the interface propagates with zero velocity. Figure 5.5 shows a plot, the solid line, of zero propagation velocity of the interface in the parameter space of k_6 and δ. Above that line the interface propagates to the right (see Fig. 5.2) and below that line to the left. For large δ (the ratio of the diffusivities) the curve of zero propagation velocity is nearly independent of δ, but for $\delta < 1$ that velocity depends on both δ and k_6.

The predictions of the thermodynamic theory presented in this chapter for equistability of two stable stationary states for the Selkov model is shown in Fig. 5.6. The results of the theory run parallel to the calculations and approach them as the length of the interface region is increased. To show this quantitatively we define the relative error

$$\text{relative error} = \frac{|\text{numerical result} - \text{theoretical result}|}{\text{numerical result}} \tag{5.52}$$

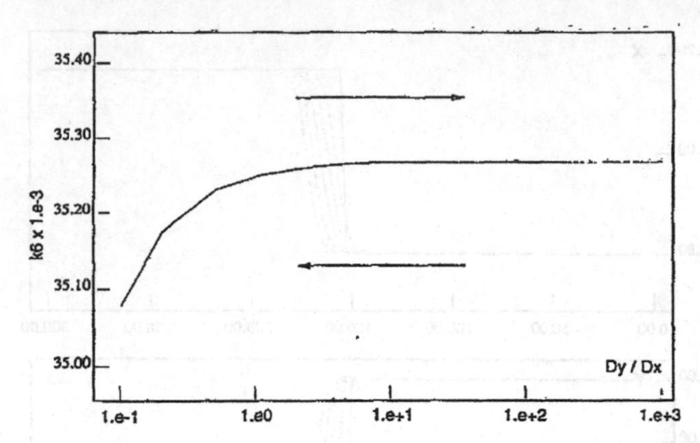

Fig. 5.5. The *solid line* is a plot of zero velocity of the interface between phase 1 and 3 calculated for the Selkov model. Above the *solid line* the interface moves to the right, below the *solid line* to the left. From [1]

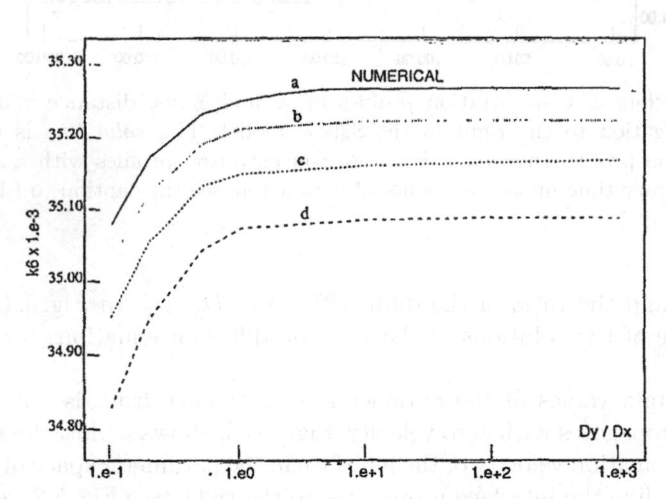

Fig. 5.6. Comparison of the predictions of equistability from the thermodynamic theory (**b**, **c**, **d**) with the numerical solution (**a**) repeated from Fig. 5.5. The theoretical results are given for different lengths of the interface region L, see Fig. 5.5: (**b**) $6L$, (**c**) $2L$, (**d**) L. The curves run parallel to the numerical calculations and approach them as the length of the interface region, and the number of boxes, are increased. From [1]

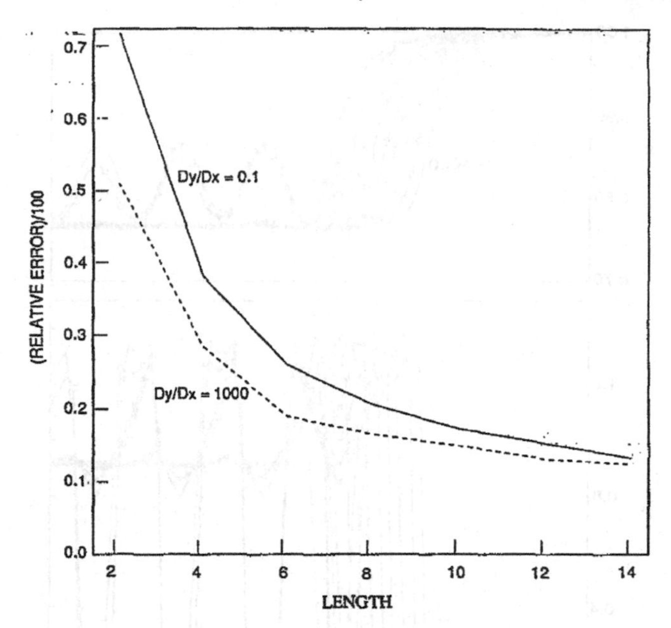

Fig. 5.7. Percent error according to (5.51) vs. the length of the interface region. See text. From [1]

and plot in Fig. 5.7 the percent error according to (5.51) vs. the length of the interface region, L, for two values of the ratio of the diffusion coefficients. The error is less than 0.2% when L exceeds 4.6×10^4, in units of length as in the diffusion coefficient D_X.

A system such as the Selkov model may have many Liapunov functions. We note that any Liapunov function of a system with multiple stationary states may serve as a criterion of relative stability [2]. Moreover the derivative of the Liapunov function with respect to L, the length of the interphase region, may also serve as such a criterion.

An interesting aside: In Fig. 5.4 we see the annihilation of a homogeneous stationary state by another homogeneous stationary state. It need not be that way always. In Fig. 5.8 we see another possible result for an enzymatic two-variable system with two stable stationary states [3] and an initial condition as in Fig. 5.4. Here, however, the interface propagates to the left but a space-dependent structure develops in the region on the right. The calculated inhomogeneous pattern is similar to that observed in dendritic solidification [4].

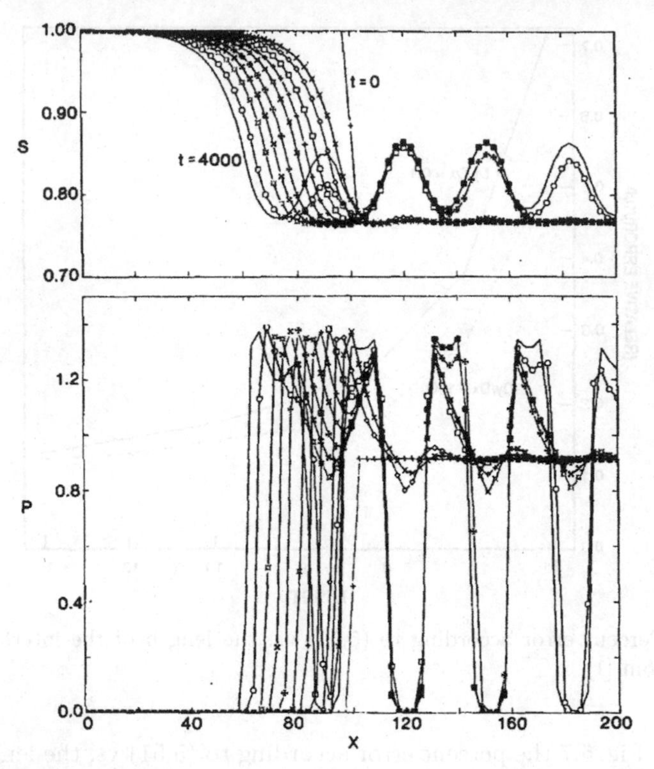

Fig. 5.8. Calculation of front propagation for a two-variable enzymatic reaction model with concentrations S and P plotted vs. distance X. The concentration profiles are given at time intervals of $\Delta t = 400$ in arbitrary units. From [3]

Acknowledgement. This chapter is based primarily on [1].

Appendix

The time derivative of the excess free energy Φ of a linear or linearized one-variable inhomogeneous system is

$$\frac{d}{dt}\Phi[\{X_i, Y_i\}] = kT \sum_{i=1}^{N} \left(\ln \frac{t_i^-}{t_i^{+\mathrm{S}}} \{t_i^+ - t_i^-\} \right), \tag{A1}$$

where

$$\begin{aligned} t_i^+ &= a_i + d(X_{i-1} + X_{i+1}) \\ &= t_i^{+S} + d(X_{i-1} - X_{i-1}^{\mathrm{S}}) + d(X_{i+1} - X_{i+1}^{\mathrm{S}}), \\ t_i^- &= (b_i + 2d)X_i = t_i^{-\mathrm{S}} + (b_i + 2d)(X_i - X_i^{\mathrm{S}}). \end{aligned} \tag{A2}$$

Substitution of (A2) into (A1) yields

$$\frac{\mathrm{d}}{\mathrm{d}t}\Phi = kT\sum_{i=1}^{N}\ln\left(1+\frac{X_i - X_i^{\mathrm{S}}}{X_i^{\mathrm{S}}}\right)[d(X_{i-1}-X_{i-1}^{\mathrm{S}}) + d(X_{i+1}-X_{i+1}^{\mathrm{S}}) - (b_i+2d)(X_i - X_i^{\mathrm{S}})]. \tag{A3}$$

With the use of the inequality

$$\ln(1+X) \leq X, \tag{A4}$$

we have

$$\begin{aligned}\frac{\mathrm{d}}{\mathrm{d}t}\Phi \leq& -kT\sum_{i=1}^{N}\frac{X_i - X_i^{\mathrm{S}}}{X_i^{S}}[d(X_{i-1}-X_{i-1}^{\mathrm{S}}) \\ &+ d(X_{i+1}-X_{i+1}^{\mathrm{S}}) - (b_i+2d)(X_i - X_i^{\mathrm{S}})] \\ =& -kT(X-X^{\mathrm{S}})^{\mathrm{T}}\mathrm{A}(X-X^{\mathrm{S}}),\end{aligned} \tag{A5}$$

where

$$(X - X^{\mathrm{S}}) = \begin{bmatrix} X_1 & - & X_1^{\mathrm{S}} \\ X_2 & - & X_2^{\mathrm{S}} \\ & \vdots & \end{bmatrix} \tag{A6}$$

and **A** is a rectangular matrix which can be written as the product of two other matrices

$$\mathbf{A} = \mathbf{BC} \tag{A7}$$

with

$$\mathbf{B} = \begin{bmatrix} 1/X_1^{\mathrm{S}} & 0 & \dots & 0 \\ 0 & 1/X_2^{\mathrm{S}} & 0 & \dots \\ & & \cdot & \\ & & & \cdot \end{bmatrix} \tag{A8}$$

and

$$\mathbf{C} = \begin{bmatrix} b_1+2d & -d & & \\ -d & b_2+2d & -d & \\ & \cdot & \cdot & \cdot \\ & & \cdot & \cdot & \cdot \end{bmatrix}. \tag{A9}$$

The matrices **B** and **C** are both positive definite and so is their product, the matrix **A**, When $(\mathbf{X} - \mathbf{X}^{\mathrm{S}})$ is not zero, the rhs of (A5) is negative definite; and when it is zero, that is at the stationary state, the rhs of (A5) is zero. Therefore the rhs of (A5) is negative semi-definite. Thus we have shown that

$$\frac{\mathrm{d}}{\mathrm{d}t}\Phi \leq 0. \tag{A10}$$

References

1. X. Chu, P.M. Hunt, K.L.C. Hunt, J. Ross, J. Chem. Phys. **99**, 3444–3454 (1993)
2. N.F. Hansen, J. Ross, J. Phys. Chem. **100**, 8040–8043 (1996)
3. H. Shyldkrot, J. Ross, J. Chem. Phys. **82**, 113–122 (1985)
4. G. Dee, J.S. Langer, Phys. Rev. Lett. **50**, 383–386 (1983)

6

Stability and Relative Stability of Multiple Stationary States Related to Fluctuations

In Chap. 5 we discussed reaction diffusion systems, obtained necessary and sufficient conditions for the existence and stability of stationary states, derived criteria of relative stability of multiple stationary states, all on the basis of deterministic kinetic equations. We began this analysis in Chap. 2 for homogeneous one-variable systems, and followed it in Chap. 3 for homogeneous multi-variable systems, but now on the basis of consideration of fluctuations. In a parallel way, we now follow the discussion of the thermodynamics of reaction diffusion equations with deterministic kinetic equations, Chap. 5, but now based on the master equation for consideration of fluctuations.

We study again the example of the Selkov model

$$A \underset{k_2}{\overset{k_1}{\rightleftharpoons}} X \quad 2Y + X \underset{k_4}{\overset{k_3}{\rightleftharpoons}} 3Y \quad Y \underset{k_6}{\overset{k_5}{\rightleftharpoons}} B, \tag{6.1}$$

which is the same as (5.49). This system may have multiple stationary states and we consider an arrangement as shown in Fig. 6.1.

As reaction and diffusion occurs an interface develops, which may travel either to the right or to the left, depending on which of the stationary states is more stable. If the two stationary states are equally stable then the velocity of the interface region is zero. We again discretize the space in the interface region and label the increments with the index i

$$L = N\Delta z \quad z_i = \mathrm{i}\Delta z \quad z_0 = 0, z_N = L, \tag{6.2}$$

where L is the length of the N segments each of width Δz. The deterministic kinetic equations are [1]

$$\begin{aligned}
\frac{\mathrm{d}X_i}{\mathrm{d}t} &= k_1 A - k_2 X_i - k_3 X_i Y_i^2 + k_4 Y_i^3 + \frac{D_X}{\Delta z^2}(X_{i+1} - 2X_i + X_{i-1}) \\
\frac{\mathrm{d}Y_i}{\mathrm{d}t} &= k_6 B - k_5 Y_i - k_3 X_i Y_i^2 + k_4 Y_i^3 + \frac{D_Y}{\Delta z^2}(Y_{i+1} - 2Y_i + Y_{i-1})
\end{aligned} \tag{6.3}$$

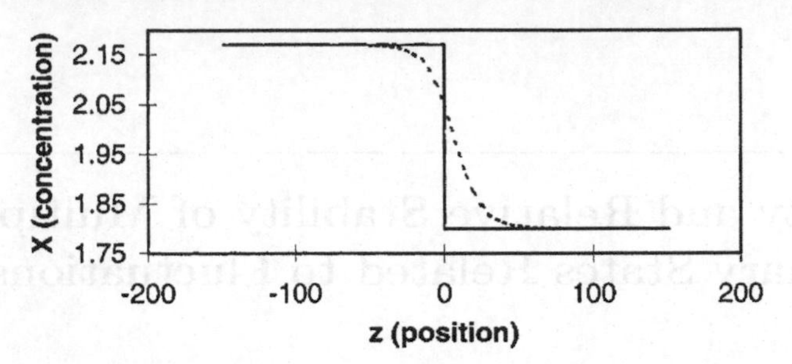

Fig. 6.1. Plot of concentration vs. position z. The initial concentration profile is shown by the solid line; the space with negative z is filled initially with stationary state 1, the space with positive z is filled initially with stationary state 3. The dotted line denotes the interphase region. Repeated from Chap. 5

written here for the discreet case. As boundary conditions we take that the concentrations of X, Y are held fixed at the left end of the system at the values of one stationary state, and at the right end at the values of the other stationary state.

We wish to consider fluctuations in concentrations of the intermediate species and proceed, as in Chap. 3, with the master equation [2]

$$\frac{\partial P\left(\overrightarrow{X},\overrightarrow{Y};t\right)}{\partial t} = \sum_{r_x,r_y}\left[W\left(\overrightarrow{X}-\overrightarrow{r}_X,\overrightarrow{Y}-\overrightarrow{r}_Y;\overrightarrow{r}_X,\overrightarrow{r}_Y\right)P\left(\overrightarrow{X},\overrightarrow{Y};t\right)\right.$$
$$\left.-W\left(\overrightarrow{X},\overrightarrow{Y};\overrightarrow{r}_X,\overrightarrow{r}_Y\right)P\left(\overrightarrow{X},\overrightarrow{Y};t\right)\right], \tag{6.4}$$

where the sum is over the elementary reactions listed in Table 6.1 for the Selkov model.

The notation is close to that in Chap. 3. The concentrations in the N boxes are denoted by vectors with arrows over the symbols. The magnitudes of the changes in X and Y are given by $\mathbf{r}$ (see text after (3.1)). We use again the eikonal approximation according to which the stationary probability of the nth stationary state is given by,

$$P^{(n)}\left(\overrightarrow{X},\overrightarrow{Y}\right) = C^{(n)}\exp\left(-S_n\left(\overrightarrow{X},\overrightarrow{Y}\right)\right),$$
$$S_n\left(\overrightarrow{X}^{(n)}_{st},\overrightarrow{Y}^{(n)}_{st}\right) = 0,$$
$$H\left(\overrightarrow{x},\overrightarrow{y};\nabla_{\overrightarrow{x}}s_n\left(\overrightarrow{x},\overrightarrow{y}\right),\nabla_{\overrightarrow{y}}s_n\left(\overrightarrow{x},\overrightarrow{y}\right)\right) = 0,$$
$$H(\overrightarrow{x},\overrightarrow{y};\overrightarrow{p}_x,\overrightarrow{p}_y) = \sum_{r_x,r_y} w(\overrightarrow{x},\overrightarrow{y};\overrightarrow{r}_x,\overrightarrow{r}_y)(\exp(\overrightarrow{r}_x\cdot\overrightarrow{p}_x+\overrightarrow{r}_y\cdot\overrightarrow{p}_y)-1),$$
$$\overrightarrow{p}_x \equiv \nabla_{\overrightarrow{x}}s_n(\overrightarrow{x},\overrightarrow{y}),\ \overrightarrow{p}_y \equiv \nabla_{\overrightarrow{y}}s_n(\overrightarrow{x},\overrightarrow{y}), \tag{6.5}$$

Table 6.1. Elementary reactions for a Sel'kov model system distributed in one dimension[a]

Elementary reaction	r_x and r_y	$W(x, y; r_x r_y)$
$A \rightarrow X_i$	$r_{x,i} = +1$	$k_i A$
$X_i \rightarrow A$	$r_{x,i} = -1$	$k_i X_i$
$X_i + 2Y_i \rightarrow 3Y_i$	$r_{x,i} = -1$	$k_i X_i Y_i^2$
	$r_{y,i} = +1$	
$3Y_1 \rightarrow X_i + 2Y_1$	$r_{x,i} = +1$	$k_i Y_i^2$
	$r_{y,i} = -1$	
$Y_i \rightarrow B$	$r_{y,i} = -1$	$k_i Y_i$
$B \rightarrow Y_i$	$r_{y,i} = +1$	$k_i B$
$X_0 \rightarrow X_1$	$r_{x,i} = +1$	$\mathrm{d}_x X_0$
$X_1 \rightarrow X_0$	$r_{x,i} = -1$	$\mathrm{d}_x X_1$
$Y_0 \rightarrow Y_1$	$r_{y,i} = +1$	$d_y Y_0$
$Y_1 \rightarrow Y_0$	$r_{y,i} = -1$	$d_y Y_1$
$X_i \rightarrow X_{i+1}$	$r_{x,i} = -1$	$d_y X_i$
	$r_{x,i+1} = +1$	
$X_{i+1} \rightarrow X_i$	$r_{x,i} = +1$	$\mathrm{d}_y X_{i+1}$
	$r_{x,i+1} = -1$	
$Y_i \rightarrow Y_{i+1}$	$r_{y,i} = -1$	$\mathrm{d}_y Y_i$
	$r_{y,i+1} = +1$	
$Y_{i+1} \rightarrow Y_i$	$r_{y,i} = +1$	$\mathrm{d}_y Y_{i+1}$
	$r_{y,i+1} = -1$	

[a] All reactions including species with an index i denote N reactions, one for each box. The terms W are the transition probabilities in the master equation, (6.4)

the first line of (6.5). The action S_n at the stationary state is zero, the second line. The Hamiltonian function is zero for the most probable fluctuational trajectory and the conjugate momentum is the gradient of the action. Next, we define the reduced quantities

$$\vec{x} \equiv \vec{X}/\Omega,\ \vec{y} \equiv \vec{Y}/\Omega,\ w(\vec{x}, \vec{y}; \vec{r}_x, \vec{r}_y) \equiv W\left(\vec{X}, \vec{Y}; \vec{r}_x, \vec{r}_y\right)\Omega \quad (6.6)$$

and

$$s_n(\vec{x}, \vec{y}) \equiv S_n\left(\vec{X}, \vec{Y}\right)/\Omega, \quad (6.7)$$

where Ω is the volume of the system, and the dot product in (6.5) has the usual vector meaning

$$\vec{f} \cdot \vec{g} \equiv \sum_{i=1}^{N} f_i g_i \quad (6.8)$$

and the sum is over all the N boxes.

The Hamiltonian equations of motion are

$$\dot{x}_i = \sum_{r_x, r_y} r_{x,i} w(\vec{x}, \vec{y}; \vec{r}_x, \vec{r}_y) \exp(\vec{r}_x \cdot \vec{p}_x + \vec{r}_y \cdot \vec{p}_y),$$

$$\dot{y}_i = \sum_{r_x, r_y} r_{x,i} w(\vec{x}, \vec{y}; \vec{r}_x, \vec{r}_y) \exp(\vec{r}_x \cdot \vec{p}_x + \vec{r}_y \cdot \vec{p}_y),$$

$$\dot{p}_{x,i} = -\sum_{r_x, r_y} (\exp(\vec{r}_x \cdot \vec{p}_x + \vec{r}_y \cdot \vec{p}_y) - 1) \nabla_{x,i} w(\vec{x}, \vec{y}; \vec{r}_x, \vec{r}_y),$$

$$\dot{p}_{y,i} = -\sum_{r_x, r_y} (\exp(\vec{r}_x \cdot \vec{p}_x + \vec{r}_y \cdot \vec{p}_y) - 1) \nabla_{x,i} w(\vec{x}, \vec{y}; \vec{r}_x, \vec{r}_y), \quad (6.9)$$

and the sum over r_x, r_y goes over all elementary reactions listed in Table 6.1. The action is obtained from the first line in (6.10)

$$\begin{aligned} s_n(\vec{x}(t), \vec{y}(t)) &= \int_{t_0}^{t} \mathrm{d}t' (\vec{p}_x \cdot \dot{\vec{x}} + \vec{p}_y \cdot \dot{\vec{y}}) \\ &= \int_{t_0}^{t} \mathrm{d}t' \sum_{r_x, r_y} (\vec{r}_x \cdot \vec{p}_x + \vec{r}_y \cdot \vec{p}_y) w(\vec{x}, \vec{y}; \vec{r}_x, \vec{r}_y) \\ &\quad \times \exp(\vec{r}_x \cdot \vec{p}_x + \vec{r}_y \cdot \vec{p}_y) \\ &= \int_{t_0}^{t} \mathrm{d}t' \sum_{r_x, r_y} \Big[(\vec{r}_x \cdot \vec{p}_x + \vec{r}_y \cdot \vec{p}_y) \\ &\quad \times \exp(\vec{r}_x \cdot \vec{p}_x + \vec{r}_y \cdot \vec{p}_y) \\ &\quad + 1 - \exp(\vec{r}_x \cdot \vec{p}_x + \vec{r}_y \cdot \vec{p}_y) \Big] w(\vec{x}, \vec{y}; \vec{r}_x, \vec{r}_y). \end{aligned} \quad (6.10)$$

We have subtracted $H(x,p) = 0$ in the third line of (6.10). We use the inequality

$$x \exp(x) + 1 - \exp(x) \geq 0$$

and the fact that w is always positive to prove that along a fluctuational path away from a stationary state we have

$$\frac{\mathrm{d}s_n(\vec{x}(t), \vec{y}(t))}{\mathrm{d}t} |_{\text{flue}} \geq 0 \, . \quad (6.11)$$

We need to prove one more important fact that we obtain from the equations

$$
\begin{aligned}
\left.\frac{\mathrm{d}s_n(\vec{x},\vec{y})}{\mathrm{d}t}\right|_{\det} &= \nabla_{\vec{x}}s_n(\vec{x},\vec{y})\left.\frac{\mathrm{d}\vec{x}}{\mathrm{d}t}\right|_{\det} + \nabla_{\vec{y}}s_n(\vec{x},\vec{y})\left.\frac{\mathrm{d}\vec{y}(\vec{x},\vec{y})}{\mathrm{d}t}\right|_{\det} \\
&= \sum_{r_x,r_y} \vec{p}_x\cdot\vec{r}_x w(\vec{x},\vec{y};\vec{r}_x,\vec{r}_y) \\
&\quad + \vec{p}_y\cdot\vec{r}_y w(\vec{x},\vec{y};\vec{r}_x,\vec{r}_y) \\
&= \sum_{r_x,r_y} (\vec{p}_x\cdot\vec{r}_x + \vec{p}_y\cdot\vec{r}_y + 1 \\
&\quad - \exp(\vec{p}_x\cdot\vec{r}_x + \vec{p}_y\cdot\vec{r}_y))w(\vec{x},\vec{y};\vec{r}_x,\vec{r}_y), \qquad (6.12)
\end{aligned}
$$

where we have made use of the relations

$$
p = \nabla s_n(x) \quad \text{and} \quad H(x,p) = 0.
$$

Finally, with the inequality

$$
x + 1 - \exp(x) \leq 0,
$$

we see that

$$
\frac{\mathrm{d}s_n(\vec{x},\vec{y})}{\mathrm{d}t}\Big|_{\det} \leq 0\,. \qquad (6.13)
$$

This proves that the stationary solution of the master equation, $s_n(x,y)$, is a Lyapunov function for the deterministic path from (x,y) to the stable stationary state.

In Chap. 3 we made the connection between the stationary solution of the master equation and the thermodynamic excess work, a state function, Φ^0. For a system with two intermediates, the variables (x,y) we can write

$$
\begin{aligned}
\mathrm{d}s &= \frac{1}{kTV}\,\mathrm{d}\phi^0 \\
&= (\mu_x - \mu_{x^0})\,\mathrm{d}n_{x,\mathrm{fl}} + (\mu_y - \mu_{y^0})\,\mathrm{d}n_{y,\mathrm{fl}}, \qquad (6.14)
\end{aligned}
$$

where (x^0, y^0) refer to a reference state given by

$$
\begin{aligned}
p_x &= \ln(x/x^0), \\
p_y &= \ln(y/y^0), \qquad (6.15)
\end{aligned}
$$

which hold for an equivalent linear system. The displacements on the rhs of (6.14) are along the most probable fluctuational path. The momentum p is the gradient of the action, and therefore ds and $\mathrm{d}\phi^0$ are exact differentials.

If we think of the interphase region divided into N boxes, then the state function of the total excess work is the sum of that in each box

$$
\begin{aligned}
\Phi^0(X,Y) &= \int_s^{(X,Y)} \sum_{i=1}^N kTV\,\mathrm{d}\phi_i^0 \\
&= kTV\sum_{i=1}^N \left[\int_s^{(X,Y)} \mathrm{d}X_i \ln\left(X_i/X_i^0\right) + \mathrm{d}Y_i \ln\left(Y_i/Y_i^0\right)\right] \qquad (6.16)
\end{aligned}
$$

The reference state is determined as in the prescription for the homogeneous case, but now in the full 2N dimensional case.

Figure 6.1 shows the developed interface region, the dotted line. At equistability this line does not move. At equistability translation of the dotted line does not change Φ^0. If the stationary state 1 (SS1) is slightly more stable than the stationary state 3 (SS3), then the deterministic motion of the front is a translation to the right. Since Φ^0 is a Lyapunov function, $\Delta\Phi^0$ for this process must be negative. Similarly for the opposite case, 3 slightly more stable than 1, $\Delta\Phi^0$ is also negative. Hence at equistability the limiting value of $\Delta\Phi^0$ for a translation along the position z must be zero.

The change in the excess work for establishing the stable front (SF) from SS1 equals at equistability the change in excess work of establishing the stable front from SS3

$$\Delta\Phi^0(\mathrm{SS1} \to \mathrm{SF}) = \Delta\Phi^0(\mathrm{SS3} \to \mathrm{SF}). \tag{6.17}$$

Thus the stationary probability distribution of the master equation in the eikonal approximation is a Lyapunov function, which gives necessary and sufficient conditions of the existence and stability of non-equilibrium stationary states and provides a measure of relative stability on the basis of inhomogeneous fluctuations, (6.17).

Acknowledgement. This chapter is based on the results in refs. [1] and [2].

References

1. N.F. Hansen, J. Ross, J. Phys. Chem. **100**, 8040 (1996)
2. N.F. Hansen, J. Ross, J. Phys. Chem. **102**, 7123 (1998)

7

Experiments on Relative Stability in Kinetic Systems with Multiple Stationary States

7.1 Multi-Variable Systems

We start with experiments on a multi-variable system, the bromate oxidation of ferroin, [1], also called the minimum bromate oscillator. The bistability and chemical oscillations of this system were characterized in [2].

The goal of these experiments is the measurement of the front propagation of one of the two stable stationary states into the other, see Chap. 5 particularly Sects. 5.1.3 and 5.1.4 and Figs. 5.2–5.7. From such measurements we can determine equistability conditions for the two stationary states where the front propagation velocity is zero. We thus obtain kinetic and thermodynamic conditions for the coexistence of the two stationary states.

Other suggestions have been made concerning the measurement of relative stability. One such suggestion [3] was based on the connections of two continuous stirred tank reactors, CSTRs, each filled with one or the other stable stationary state; the final state of both CSTR is predicted to be the more stable stationary state. However, the final state has been shown theoretically [4] and experimentally [5,6] to depend also on the strength and manner of mixing of the CSTRs, and therefore is not a useful, direct measure of relative stability.

We assume that the reaction mechanism of this reaction given by Noyes et al. (NFT) [7] is adequate, although recognized to be perhaps over-simplified:

$$
\begin{gathered}
\mathrm{BrO_3{}^- + Br^- + 2H^+ \rightarrow HOBr + HBrO_2},\\
\mathrm{HBrO_2 + Br^- + H^+ \rightarrow 2HOBr},\\
\mathrm{HOBr + Br^- + H^+ \rightarrow Br_2 + H_2O},\\
\mathrm{HBrO_2 + BrO_3{}^- + H^+ \rightarrow 2BrO_2^{\cdot} + H_2O},\\
\mathrm{BrO_2^{\cdot} + Fe(phen)_3{}^{2+} + H^+ \rightarrow Fe(phen)_3{}^{3+} + HBrO_2},\\
\mathrm{HBrO_2 \rightarrow HOBr + BrO_3{}^- + H^+}.
\end{gathered}
$$

We measured bistability (1) for this reaction and the results are shown in Fig. 7.1.

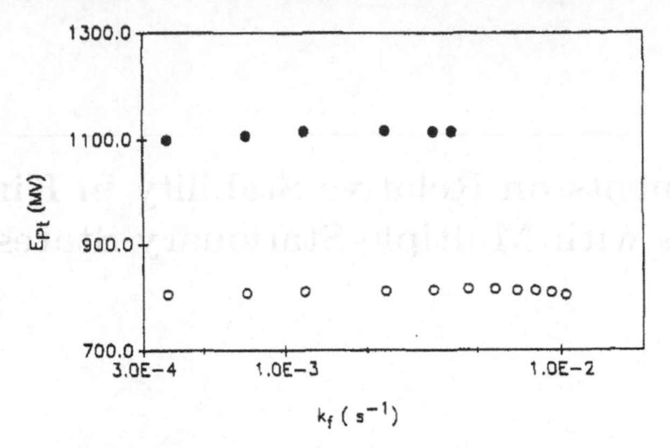

Fig. 7.1. Measured hysteresis in our reaction system [1]. Plot of the redox potential of Br^-, E_{Pt}, as a function of the flow rate coefficient k_f (in units of reciprocal residence times, the time spent by a volume element in the laminar flow reactor (LFR)). Filled dots represent one of the stable stationary states (the oxidized state) and empty dots the other stable state, the reduced state. From [1]

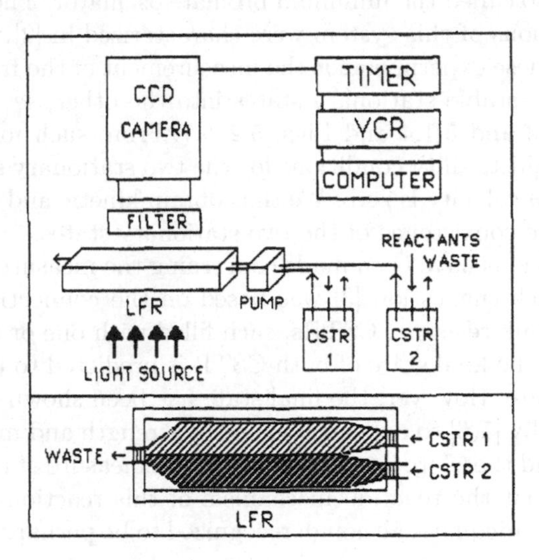

Fig. 7.2. Schematic diagram of the apparatus. Each solution, one corresponding to one stable stationary state and the other to the other stationary state, is stored in one of two continuous-stirred tank reactors (CSTR) and pumped at a determined and variable rates through the laminar flow reactor (LFR), where they are brought in contact with each other in a sharp well-defined boundary. For the remainder of the definitions see the text. From [1]

For an apparatus we choose one to be related to the schematic versions given in Chap. 5. Two solution mixtures, one corresponding to one of the stable stationary states and the other to the other stationary state, are flowed through a laminar flow reactor [1], Fig. 7.2, in two very thin (1 mm) streams

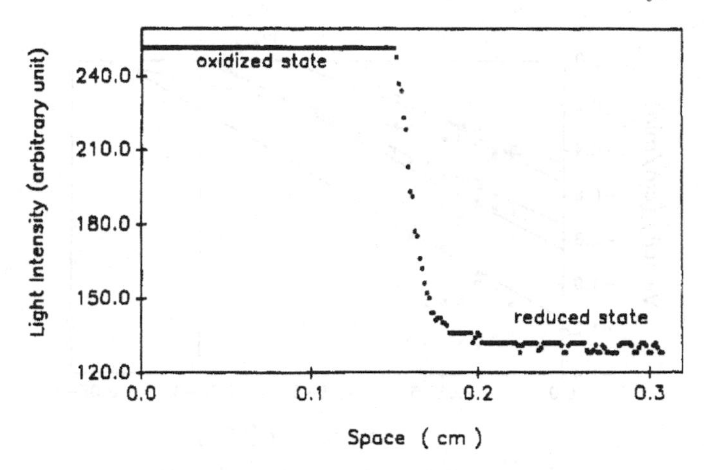

Fig. 7.3. Intensity profile across the interface of the two stationary states brought in contact in the laminar flow reactor. Both solutions flowed through the reactor at a rate such that there is no time for diffusion to occur. This measurement is taken prior to the start of the experiment itself. From [1]

in contact with each other on one edge of each solution. There is hardly any mixing due to flow, see Fig. 7.3, which shows measurements of the sharp interface, of extent less than 0.5 mm, between the two layers. In these experiments the blue (oxidized) stationary state is more stable than the reduced state. After a transient period, to establish stationary conditions, the flows of the two streams are stopped and diffusion occurs across the interface, since the two stationary states have different concentrations of the various species. Two time scales are of importance here: the rate of front propagation of one solution into the other, and the rate of change of concentrations in the stopped solutions. From various estimates we determined that the concentrations of the solutions due to homogeneous chemical reactions in each of the streams changed but little in about 60 s. Hence measurements of front propagation of one solution into the other was restricted to such times. The measurements were made by light absorption with an interference filter (Corion) and an absorption maximum at 490 nm that converted the red/blue image to a contrast enhanced black/white image. The light intensity is measured and recorded in a charge coupling device (CCD) and then processed by computer.

The experimental results are shown in Fig. 7.4.

In an experiment we measured the front propagation five times for each flow rate. The five velocities at each flow rate are averaged and a symbol in Fig. 7.4 represents that average. The experimental points are then fitted to a straight line and extrapolated to zero velocity, which gives the flow rate for equistability. The average value of the flow rate at equistability determined from the measurements is $(6.1 \pm 0.6) \times 10^{-3}\,\mathrm{s}^{-1}$.

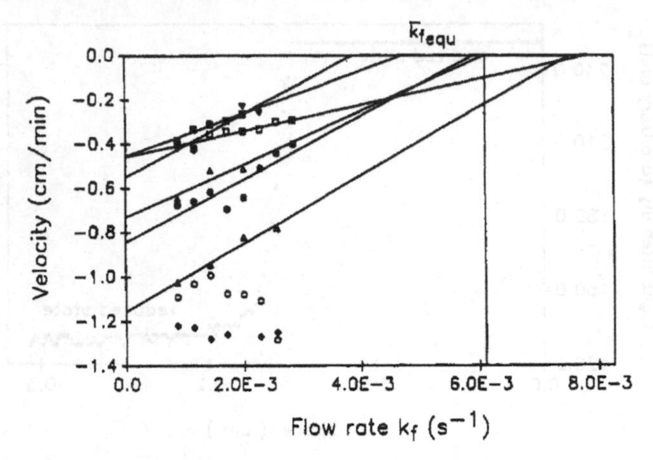

Fig. 7.4. Plots of the measured dependence of the velocity of front propagation of one stationary state into the other on the flow rate, k_f. The eight different symbols correspond to eight experiments done under the same conditions on different days. Two of the plots are shifted from their original positions for the purpose of better display: circles by $-0.6\,\mathrm{cm\,min^{-1}}$ in V; diamonds by $-0.7\,\mathrm{cm\,min^{-1}}$ in V. Lines are fitted to each set of points for purpose of extrapolation to zero velocity of front propagation. The precision of the points at the largest velocities is insufficient to permit extrapolation. The average value of the flow rate coefficient at zero velocity of propagation is $6.1 \pm 0.6 \times 10^{-3}\,\mathrm{s^{-1}}$. From [1]

This experimental result was compared with a calculation based on the NFT mechanism R1–R6, yet further simplified to a two-variable system [6]. Deterministic reaction–diffusion equations were solved numerically as described in Sect. 5.1.4, and the value of the flow rate at equistability was determined to be $12.2 \times 10^{-3}\,\mathrm{s^{-1}}$.

A further comparison based on the thermodynamic theory of reaction diffusion systems presented in Sects. 5.1.3 and 5.1.4, yields the value of $12.45 \times 10^{-3}\,\mathrm{s^{-1}}$. In view of the difficulties and limited precision of the experiments and the use of a very simple model of the reaction mechanism the agreements of the experiment with the thermodynamic theory and the calculations are satisfactory.

For more details on the calculations and the application of the thermodynamic theory for this particular reaction see [1].

7.2 Single-Variable Systems: Experiments on Optical Bistability

Interference filters can be made with ZnSe and alternating layers of ThF_4 in a stack. Irradiation of such filters with light from an argon laser (514 nm) produces optical bistability in certain ranges of power of the irradiating light:

a plot of the power of the light transmitted by the filter vs. the power of the irradiating light shows hysteresis and hence multiple stationary states. A part of the incoming light is absorbed by the filter and turned into heat. The resulting increase in temperature changes the band gap in the semiconductor, which alters the absorption. This feedback leads to multiple stationary states; the equation of change for the temperature is [8]

$$\frac{\mathrm{d}T}{\mathrm{d}t} = \frac{P_{\mathrm{in}}\mathrm{Abs}[T]}{CdA} - \frac{T - T^0}{\tau}, \tag{7.1}$$

where P_{in} is the wattage of the irradiating light, d is the depth of the filter, A is the area of the irradiated surface of the filter, $\mathrm{Abs}[T]$ is the calculated fraction of the light absorbed, T^0 is the room temperature (the temperature of the filter not irradiated), and τ is the combined heat transfer coefficient for convection, radiation and conduction.

On irradiation of a given region of the filter there results two possible temperature profiles [8] as given by the transmitted light, or the absorbed light, as a function of the distance z, (calculated) see Fig. 7.5.

In Fig. 7.5a the power of the incident light onto the filter is sufficient to keep most of the irradiated region in the upper (higher T) of the two possible stationary states labelled with T_3. The non-irradiated region is at T^0 (about 20°C). There is a short connecting region in the lower stationary state,

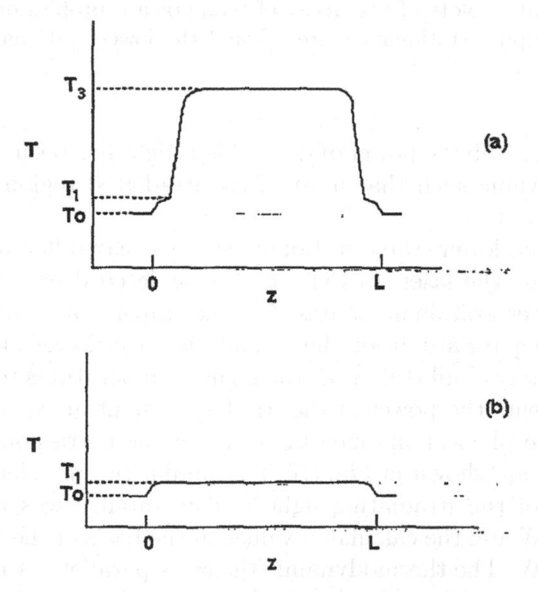

Fig. 7.5. Two possible stable stationary states of an optically bistable interference filter. The region of irradiation is the length L; the ambient temperature is T^0; the upper temperature is T_3 and the lower one is T_1. From [8]

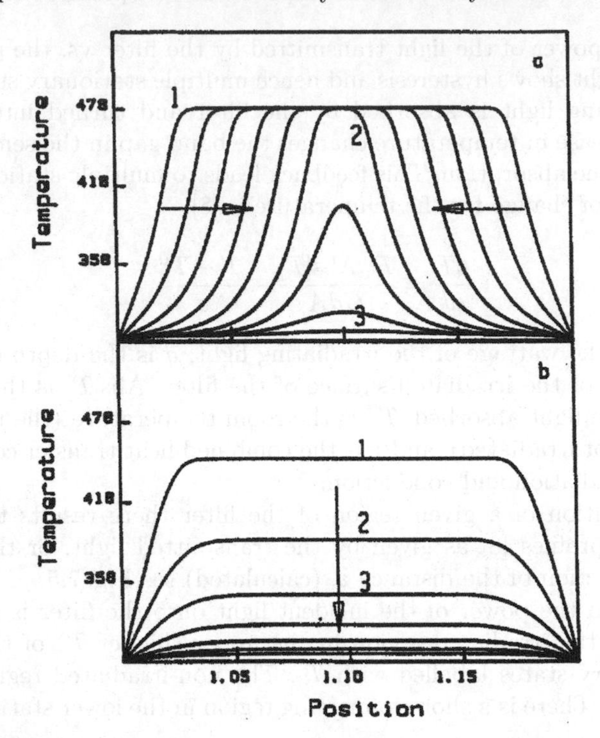

Fig. 7.6. Calculated plots of the decay of temperature profiles on stoppage of irradiation of the upper stationary state (**a**) and the lower stationary state (**b**); see text. From [9]

labelled 1. In Fig. 7.5b the power of the incident light has been increased slowly from zero to a value such that most of the irradiated region is in the lower stable stationary state.

The upper and lower stable stationary states decay differently on stopping the irradiation by the laser. In Fig. 7.6 we calculated decays from the upper (a) and the lower stationary states [9]. The upper state (a) is annihilated by fronts moving inward from the boundaries; the lower state (b) by the simultaneous decay of all regions at the higher temperatures toward T^0. From such measurements the power of the irradiating light at equistability can be determined. The plots of the measurements of the corresponding calculated plots in Fig. 7.6 are shown in Fig. 7.7. The similarities are clear.

The power of the irradiating light at equistability was measured to be 1.395 ± 0.015 mW and the calculated value, on the basis of the thermodynamic theory, 1.375 mW. The thermodynamic theory is parallel of that for reaction–diffusion but amended for a problem in thermal conduction [8]. The agreement may be better than warranted.

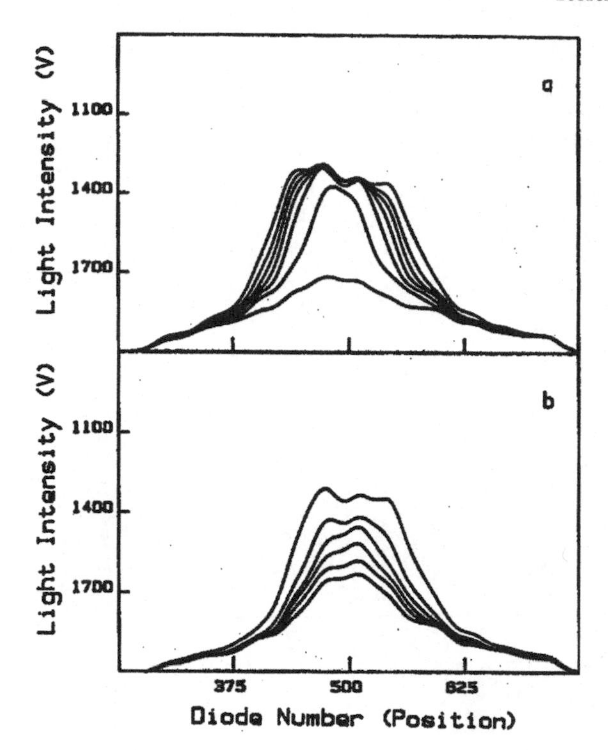

Fig. 7.7. Measured plots of the decay of temperature profiles on stoppage of irradiation of the upper stationary state (**a**) and the lower stationary state (**b**); see text. From [9]

Acknowledgement. This chapter is based on refs. [1,8,9].

References

1. P. Foerster, Y.-X. Zhang, J. Ross, J. Phys. Chem. **97**, 4708 (1993); The reaction mechanism given in this ref, eqs. R1–R6 has a misprint in R5, corrected in this chapter
2. V. Gáspár, G. Bazsa, M.T. Beck, J. Phys. Chem. **89**, 5495 (1985)
3. R.M. Noyes, J. Chem. Phys. **71**, 5144 (1979)
4. K. Hunt, J. Kottalam, M. Hatlee, J. Ross, J. Chem Phys. **96**, 7019 (1992)
5. I. Stuchl, M. Marek, J. Chem. Phys. **77**, 1607 (1982)
6. K. Bar-Eli, W. Geiseler, J. Phys. Chem. **85**, 3461 (1981)
7. R.M. Noyes, R.J. Field, R.C. Thompson, J. Am. Chem. Soc. **93**, 7315 (1971)
8. A.N. Wolff, A. Hjelmfelt, J. Ross, J. Chem. Phys. **99**, 3455 (1993)
9. R.H. Harding, J. Ross, J. Chem. Phys. **92**, 1936 (1990)

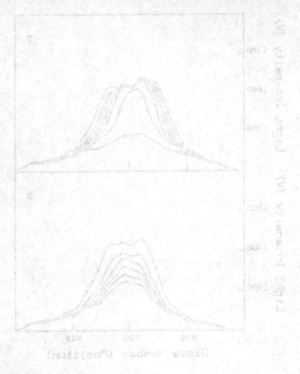

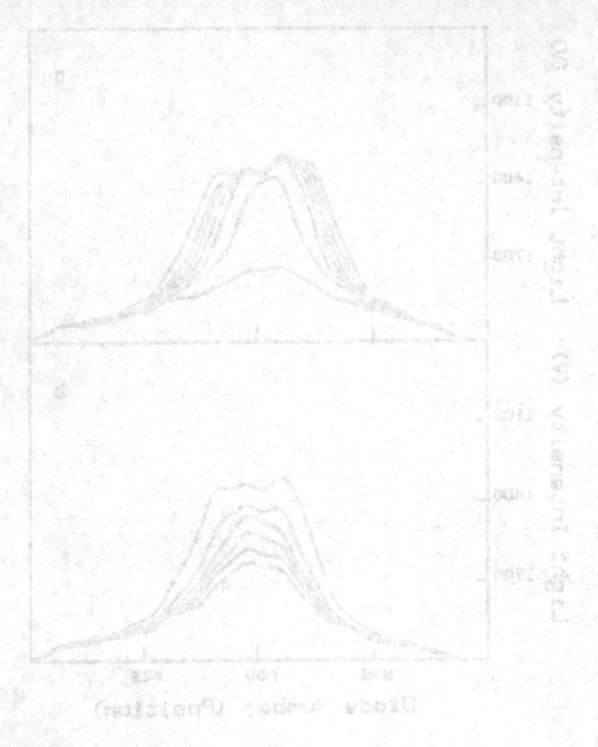

8

Thermodynamic and Stochastic Theory of Transport Processes

8.1 Introduction

In this chapter we present the thermodynamic and stochastic theory of simple transport processes, linear and non-linear: diffusion, thermal conduction and viscous flow. We refer only briefly to more advanced work on hydrodynamic equations and some interesting experiments, and on light scattering from a fluid in a temperature gradient. A suitable introduction to the chapter may well be a review of the some of the main results obtained in earlier chapters on chemical reactions, and we begin with that review.

For a model chemical reaction such as the Schlögl model

$$A + 2X \underset{k_2}{\overset{k_1}{\Longleftrightarrow}} 3X$$

$$X \underset{k_4}{\overset{k_3}{\Longleftrightarrow}} B, \tag{8.1}$$

we define a species-specific affinity

$$\Phi(X) = \int^x (\mu_x - \mu_{x^*})\, \mathrm{d}X, \tag{8.2}$$

which is an excess work relative to a starred reference state defined by the kinetics of the system. The relation of (8.2) to the chemical kinetics comes from the rate equation for the intermediate X

$$\mathrm{d}X/\mathrm{d}t = t_x{}^+ - t_x{}^-, \tag{8.3}$$

where

$$\begin{aligned} t_x{}^+ &= k_1 A X^2 + k_4 B \\ t_x{}^- &= k_2 X^3 + k_3 X \end{aligned} \tag{8.4}$$

for the model (8.1). The relation is

$$\mu_x - \mu_{x^*} = -kT \ln\left(\frac{t_x{}^+}{t_x{}^-}\right) \tag{8.5}$$

and the quantity μ_{X^*} approaches its value in each of the stationary states μ_X^S as each of the stationary states in the reaction model is approached. The theory reviewed here can be generalized to multi-variable systems and to other choices of reference states (see Chap. 3).

From (8.2) we can draw several important consequences:

1. The function Φ provides necessary and sufficient conditions of global stability; it has an extremum at each stationary state

$$\partial\Phi/\partial X = 0 \tag{8.6}$$

and its second derivative shows whether a given stationary state is stable or unstable

$$\frac{\partial^2\Phi}{\partial X^2}\begin{cases} > 0 & \text{for stable stationary states} \\ < 0 & \text{for unstable stationary states.} \end{cases} \tag{8.7}$$

2. The function Φ is the driving force (frequently so-called, in fact a potential) toward stable stationary states. Φ is lower bounded and is a Liapunov function with the derivative of Φ with respect to time

$$\dot{\Phi} = -kT\,\ln\left(\frac{t_x{}^+}{t_x{}^-}\right)[t_x{}^+ - t_x{}^-] \leq 0 \tag{8.8}$$

in the form of Boltzmann's H theorem.
3. $\dot{\Phi}$ is a component of the total dissipation and is zero at stationary states.
4. $\mathrm{d}\Phi$ is a measureable excess work: for a linear system it is the negative of the work, other than pressure–volume work, obtained from the spontaneous relaxation $\mathrm{d}X$ at X along the deterministic trajectory, minus the work of displacing the system the same extent $\mathrm{d}X$ at X_{S}. For a non-linear system we substitute in the prior sentence X^* for X_{S}. For each X of a non-linear system there is an X^* and an instantaneous linear equivalent system at each instant, which makes this substitution possible.
5. If a system approaches the stationary state of equilibrium then Φ approaches correctly the thermodynamic free energy change.
6. The function Φ provides a criterion of relative stability of two stable stationary states, labelled 1 and 3; at equistability of the two stationary states the condition holds

$$\int_1^3 (\mu_x - \mu_{x^*})\,\mathrm{d}X = 0. \tag{8.9}$$

7. The function Φ yields the stationary probability distribution of a stochastic, birth–death master equation for single variable systems

$$P_{\mathrm{st}}(X) = P_0\,\exp\left(-\frac{\Phi(X)}{kT}\right), \tag{8.10}$$

where P_0 is a normalization constant. The combination of (8.2) and (8.10) is a generalization of the fluctuation equation of Einstein to stationary non-equilibrium states and conditions far from equilibrium. For one-variable systems the excess work Φ is a state function. In general the excess work is not a state function for the reference state as defined in this section (see also Chap. 4). The excess work is a state function for the reference state as defined in Chap. 3.

8.2 Linear Transport Processes

In this chapter we formulate the thermodynamic and stochastic theory of the simple transport phenomena: diffusion, thermal conduction and viscous flow (1) to present results parallel to those listed in points 1–7, Sect. 8.1, for chemical kinetics. We still assume local equilibrium with respect to translational and internal degrees of freedom. We do not assume conditions close to chemical or hydrodynamic equilibrium. For chemical reactions and diffusion the macroscopic equations for a given reaction mechanism provide sufficient detail, the fluxes in the forward and reverse direction, to write a birth–death master equation with a stationary solution given in terms of Φ. For thermal conduction and viscous flow we derive the excess work Φ and then find Fokker–Planck equations with stationary solutions given in terms of that excess work.

8.2.1 Linear Diffusion

Consider the simple experiment shown schematically in Fig. 8.1.

The total pressure is essentially constant, because of the large excess of A, but diffusion of X occurs because we set the number of molecules of X in volume 1, X_1, larger than X_3.

We assume that the diffusion of X across the membranes separating volumes 1 and 2, and 2 and 3 is slow compared to the time necessary to achieve a homogeneous concentration of X in volume 2 by diffusion or stirring. The macroscopic temporal variation of the number of X in volume 2 (V), with symbol X, is

$$\mathrm{d}X/\mathrm{d}t = V(k_1 x_1 + k_3 x_3) - (k_1 + k_3)X = t^+ - t^- , \tag{8.11}$$

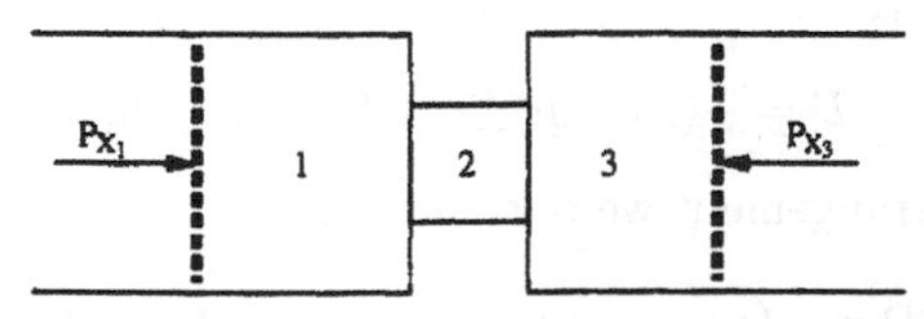

Fig. 8.1. Two species A and X are present: A is distributed homogeneously in all three volumes 1, 2, 3, in large excess over species X. Two pistons with membranes permeable to A but not to X keep the concentration of X in volume 1 different from that in volume 3, so that diffusion occurs

where the concentration of X is x, and the stationary concentration of X is given by

$$X_{\mathrm{s}} = \frac{V(k_1 x_1 + k_3 x_3)}{k_1 + k_3}. \tag{8.12}$$

We write the chemical potential of X, considered an ideal gas, as

$$\mu(X,T) = \mu^{\circ}(T) + kT \ln X, \tag{8.13}$$

where the first term on the rhs is the standard chemical potential. Hence we have for the difference in the chemical potential of X and that in the stationary state

$$\mu - \mu_{\mathrm{s}} = kT \ln\left(\frac{X}{X_{\mathrm{s}}}\right) = kT \ln\left(\frac{t^-}{t^{+\mathrm{S}}}\right). \tag{8.14}$$

In this equation we have

$$t^{+\mathrm{S}} = t^+ = V(k_1 x_1 + k_3 x_3). \tag{8.15}$$

We now define the function Φ, an excess work, as

$$\Phi = \int^x (\mu - \mu_{\mathrm{s}})\mathrm{d}X = kT \int^x \ln\left(\frac{t^-}{t^{+\mathrm{S}}}\right)\mathrm{d}X \tag{8.16}$$

in analogy with (8.2). At the stationary state we have

$$(\partial\Phi/\partial X)_{\mathrm{st}} = 0 \tag{8.17}$$

and Φ is therefore an extremum at that state. The second derivative of Φ with respect to X at that state

$$\partial^2\Phi/\partial X^2 = kT/X > 0 \tag{8.18}$$

is positive definite and hence the stationary state is stable. Φ is a Liapunov function; it is bounded, that is $\Phi \geq 0$, and $\dot{\Phi}$ is

$$\dot{\Phi} = -kT \ln\left(\frac{t^+}{t^-}\right)[t^+ - t^-] \leq 0. \tag{8.19}$$

All the equations are valid for arbitrarily large displacement from diffusional equilibrium. The driving force for evolution towards the stationary state, (8.14), is not linearly related to the flux of that evolution, (8.11).

The dissipation is

$$\mathcal{D} = -(\mu_1 - \mu)\dot{X}_1 + (\mu - \mu_3)\dot{X}_3, \tag{8.20}$$

and on further rearrangement we obtain

$$\mathcal{D} = -(\mu_1 - \mu_{\mathrm{s}})\dot{X}_1 + (\mu_{\mathrm{s}} - \mu_3)\dot{X}_3 - \dot{\Phi}. \tag{8.21}$$

The total dissipation is the sum of the dissipation in the stationary state, the first two terms in (8.21) and $-\dot{\Phi}$.

We assume that the fluctuations in this system are given by a birth–death master equation

$$\begin{aligned} \mathrm{d}P(X,t)/\mathrm{d}t = &V_2(k_1x_1 + k_3x_3)P(X-1,t) + (k_1+k_3)(X+1)P(X+1,t) \\ &- \{V_2(k_1x_1 + k_3x_3) + (k_1+k_3)X\}P(X,t) \end{aligned} \tag{8.22}$$

for which the stationary (time-independent) solution in the thermodynamic limit is

$$P_\mathrm{S}(X) = P_0 \exp\left(\int^X \ln\left(\frac{t^+(X')}{t^-(X')}\right)\mathrm{d}X'\right) = P_0 \exp\left(-\frac{1}{kT}\Phi\right). \tag{8.23}$$

Compare with (2.29, 2.33 and 2.34) in Chap. 2 for chemical systems. The macroscopic driving force towards the stationary state, (8.16), also determines the fluctuations from the stationary state, (8.23).

The generalization to systems with spatial variations, see Fig. 5.1, has been discussed in Chap. 5.

8.2.2 Linear Thermal Conduction

We follow a path for this transport process analogous to diffusion and chemical reaction. Consider a simple schematic of an apparatus consisting of two heat reservoirs, each of infinite heat capacity, one at temperature T_1 and the other at T_2, with $T_1 > T_2$, Fig. 8.2.

Volume 2 is between two thermal reservoirs labelled 1 and 3. Volume 2 is of small width so that its temperature T is uniform within it. The flow of heat occurs with conservation of energy and no work done.

We write for the 'mixed' thermodynamic function M

$$\mathrm{d}M = \mathrm{d}S_1 + \mathrm{d}S + \mathrm{d}S_3, \tag{8.24}$$

where S is the entropy, or

$$\mathrm{d}M = \frac{\mathrm{d}Q_1}{T_1} + \frac{\mathrm{d}Q}{T} + \frac{\mathrm{d}Q_3}{T_3}. \tag{8.25}$$

At the stationary state we have

$$\mathrm{d}M_\mathrm{S} = \frac{\mathrm{d}Q_1}{T_1} + \frac{\mathrm{d}Q}{T_\mathrm{S}} + \frac{\mathrm{d}Q_3}{T_3}. \tag{8.26}$$

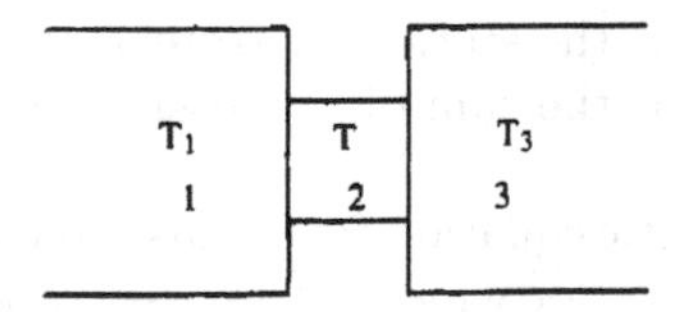

Fig. 8.2. Schematic apparatus for study of thermal conduction

Hence the driving force towards the stationary state is

$$T_S^{-1}\,d\Phi = -(dM - dM_S) = \left(\frac{1}{T_S} - \frac{1}{T}\right) dQ \tag{8.27}$$

for the same changes dQ_1, dQ, dQ_3. The integral of $d\Phi$ is

$$\Phi(T) = \Phi(T) - \Phi(T_S) = C_V\left(T - T_S - T_S \ln\left(\frac{T}{T_S}\right)\right). \tag{8.28}$$

This is an excess work as seen from the last two equations.

The macroscopic transport equation for this thermal conduction process is

$$dT/dt = \kappa_1(T_1 - T) + \kappa_3(T_3 - T) = F(T), \tag{8.29}$$

where κ_1 and κ_3 are proportional to thermal conduction coefficients for the interface of the system with the heat reservoirs 1 and 3, respectively. The temperature of the stationary state is

$$T_S = \frac{\kappa_1 T_1 + \kappa_3 T_3}{\kappa_1 + \kappa_3} \tag{8.30}$$

and hence we may write

$$dT/dt = (\kappa_1 + \kappa_3)(T_S - T). \tag{8.31}$$

In (8.27) we divide by dt, use the equation $\frac{dQ}{dt} = C_V\,dT$ and obtain for the time derivative of Φ

$$\dot{\Phi} = C_V\left(1 - \frac{T_S}{T}\right) 2\kappa(T_S - T) \le 0; \quad 2\kappa = \kappa_1 + \kappa_3. \tag{8.32}$$

Since $\Phi(T) \ge 0$ and lower-bounded we see that it is a Liapunov function.

The derivative of $\Phi(T)$ with respect to T is

$$\frac{d\Phi}{dT} = C_V\left(1 - \frac{T_S}{T}\right)\begin{cases} > 0 & T > T_S \\ = 0 & T = T_S \\ < 0 & T < T_S \end{cases} \tag{8.33}$$

and the second derivative with respect to T is

$$\frac{d^2\Phi}{dT^2} = \frac{C_V}{T^2} \ge 0 \tag{8.34}$$

so that Φ is a minimum at the stationary state T_S.

The total dissipation is the sum of the dissipation of the reservoirs in the stationary state, and $\dot{\Phi}$.

Next we seek a stochastic equation for the distribution of fluctuations of the macroscopic temperature T for which the excess work (8.28) is its stationary distribution. This is a Fokker–Planck type equation

$$\frac{\partial}{\partial t}P(T,t) = -\frac{\partial}{\partial T}(F(T)\ P(T,t)) + \frac{\partial^2}{\partial T^2}\{(f(T)\ P(T,t))\} \tag{8.35}$$

in which the probability diffusion coefficient f(T) is cT with $c = k(\kappa_1 T_1 + \kappa_3 T_3)/C_\mathrm{V}$, which is constant. The stationary solution of the stochastic equation is

$$P_\mathrm{S}(T) = P_0 \frac{T_\mathrm{S}}{T} \exp\left(-\frac{1}{kT_\mathrm{S}}\Phi\right). \tag{8.36}$$

We rescale the temperature $\tau = C_\mathrm{V}/kT_\mathrm{S}T$ and use Stirling's approximation for $\ln X! = X \ln X - X$ and obtain for the stationary distribution

$$P_\mathrm{S}(\tau) = \frac{P_0'}{\tau} \exp\left(-\tau + \tau_\mathrm{S} + \tau_\mathrm{S} \ln\left(\frac{\tau}{\tau_\mathrm{S}}\right)\right) \approx \frac{\tilde{P}_0}{\Gamma(\tau_\mathrm{S})} \tau^{\tau_\mathrm{S}-1}\, e^{-\tau}, \tag{8.37}$$

where Γ is the gamma function. If we expand this solution around T_S we arrive at the expected quadratic form

$$\Phi(T) = C_\mathrm{V}\left(T - T_\mathrm{S} - T_\mathrm{S} \ln\left(\frac{T}{T_\mathrm{S}}\right)\right) \approx \frac{C_\mathrm{V}}{2}\frac{(T-T_\mathrm{S})^2}{T_\mathrm{S}}. \tag{8.38}$$

and the Gaussian stationary distribution

$$P_\mathrm{S}(T) = P_0 \exp\left(-\frac{\Phi}{kT_\mathrm{S}}\right) = P_0 \exp\left(-\frac{C_\mathrm{V}}{2k{T_\mathrm{S}}^2}(T-T_\mathrm{S})^2\right). \tag{8.39}$$

This is exactly the same form as the equilibrium probability distribution for the fluctuations in temperature (2) at the equilibrium temperature $T_\mathrm{S} = T_\mathrm{equ}$. For an ideal gas we have $E = C_\mathrm{V}T$ and the fluctuations are in the Gaussian form in energy

$$\Phi(E) = \frac{1}{2}\frac{(E-E_\mathrm{S})^2}{C_\mathrm{V}T_\mathrm{S}}. \tag{8.40}$$

The generalization to a system with a discreet distribution of temperature follows closely the development for diffusion.

For an experimental test of the theory for thermal conduction see Sect. 7.2.

8.2.3 Linear Viscous Flow

We consider both cases of simplified Couette flow, Fig. 8.3, and Poiseuille flow, Fig. 8.4, see [3]

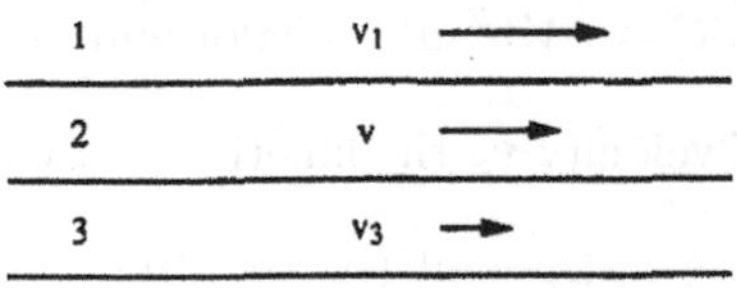

Fig. 8.3. Simple model for Couette flow. Section 1 and 3 are moving with fixed but different speeds. Section 2 relaxes to a stationary state through frictional forces on the interfaces between 1 and 2, and 2 and 3

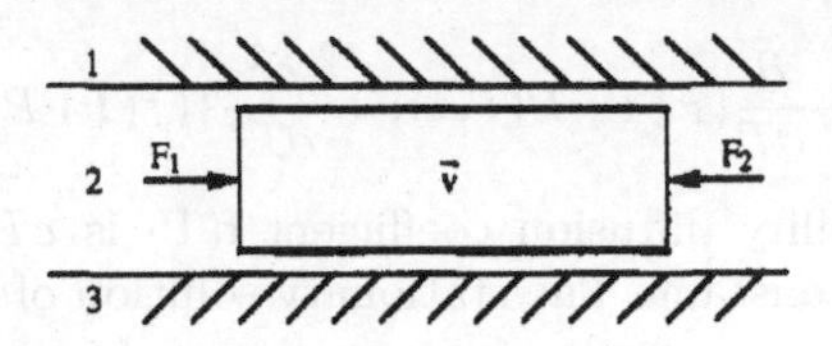

Fig. 8.4. Simple model for Poiseuille flow. A block of fluid, gas or liquid, is pushed through a narrow region between neighbouring blocks at rest

The Newtonian equation for Couette flow is

$$\rho \frac{\mathrm{d}v}{\mathrm{d}t} = \alpha(v_1 - v) - \alpha(v - v_3) = 2\alpha(v_S - v), \tag{8.41}$$

where α is the friction coefficient at the interfaces between the system and the two neighbouring blocks. The steady state mass velocity is $\nu_S = 1/2(\nu_1 + \nu_3)$ and ρ is the mass density of the system. Because of the shear forces on the system there is a generation of heat at the interfaces of region 1 and 2, and of regions 2 and 3. These dissipative terms are usually neglected in hydrodynamics but are essential here. We assume that the entire system, regions 1, 2, and 3, are at one temperature.

For Poiseuille flow the middle block is driven by constant force $fV = F_1 - F_2$ with f the force density and V the volume of the system. Newton's equation is

$$\rho \frac{\mathrm{d}v}{\mathrm{d}t} = f - 2\alpha v = 2\alpha(v_S - v), \tag{8.42}$$

which is the same as that for Couette flow.

In the case of Coutte flow there is no pressure–volume work

$$\mathrm{d}W^{\mathrm{pv}} = -p\,\mathrm{d}V = 0. \tag{8.43}$$

But in the case of Poiseuille flow the pressure–volume work done on the system is

$$\mathrm{d}W = -(p_1 - p_2)\,\mathrm{d}V = -(p_1 - p_2)\mathrm{S}v\,\mathrm{d}t = -fVv\,\mathrm{d}t = -2\alpha V v_S v\,\mathrm{d}t, \tag{8.44}$$

where S is the surface over which the force is exerted and p is the pressure. The heat generated by dissipation (friction) is in the two cases

$$\begin{aligned} \mathrm{d}Q &= {}^1\!/\!_2\alpha V((v_1 - v)^2 + (v_3 - v)^2)\mathrm{d}t \qquad \text{(Couette)}, \\ \mathrm{d}Q &= \alpha V v^2\,\mathrm{d}t \qquad \text{(Poiseuille)}. \end{aligned} \tag{8.45}$$

In the moving frame of velocity v_S the kinetic energy changes $\mathrm{d}K$ are

$$\mathrm{d}K = \rho V(v - v_S)\mathrm{d}v \tag{8.46}$$

in both types of flow.

The minimum work necessary for removing the system from a stationary state is given by considering changes in kinetic energy. For the three sections these are

$$\mathrm{d}E_1 = v_1\,\mathrm{d}P_1, \quad \mathrm{d}E_2 = v_2\,\mathrm{d}P_2, \quad \mathrm{d}E_3 = v_3\,\mathrm{d}P_3, \tag{8.47}$$

where $\mathrm{d}P_i$ is the momentum of the ith section. The composite of $\mathrm{d}M$ equals $\mathrm{d}E_1+\mathrm{d}E_2+\mathrm{d}E_3$, and hence for the differential minimum excess work we have

$$\mathrm{d}\Phi = \mathrm{d}M - \mathrm{d}M_\mathrm{S} = (v - v_\mathrm{S})\mathrm{d}P = \rho V(v - v_\mathrm{S})\,\mathrm{d}v, \tag{8.48}$$

where we omit the subscript 2, and

$$\mathrm{d}M_\mathrm{S} = v_1\,\mathrm{d}P_1 + v_{2\mathrm{S}}\,\mathrm{d}P_2 + v_3\,\mathrm{d}P_3. \tag{8.49}$$

For the properties of the excess work we find

$$\Phi = {}^1\!/_2 \rho V(v - v_\mathrm{S})^2 \geq 0, \tag{8.50}$$

where the equality in the second equation holds only at the stationary state. The time derivative of Φ, with the use of (8.42) is

$$\dot{\Phi} = -2\alpha V(v - v_\mathrm{S})^2 \leq 0 \tag{8.51}$$

and again the equality in the second equation holds only at the stationary state. The last two equations show that Φ is a Liapunov function. The excess work is the kinetic energy of section 2 relative to a coordinate system moving with velocity v_S.

At the stationary state we have the necessary and sufficient extremum condition

$$\mathrm{d}\Phi/\mathrm{d}v = 0, \tag{8.52}$$

and the stationary state is stable since

$$\mathrm{d}^2\Phi/\mathrm{d}v^2 = \rho V > 0. \tag{8.53}$$

For Couette flow the total dissipation is

$$\frac{\mathrm{d}Q}{\mathrm{d}t} = \alpha V((v_1 - v)^2 + (v_3 - v)^2) = \frac{\mathrm{d}Q_\mathrm{S}}{\mathrm{d}t} - \frac{\mathrm{d}\Phi}{\mathrm{d}t} \tag{8.54}$$

and we may write

$$D_\mathrm{total} = D_\mathrm{st} - \dot{\Phi} \tag{8.55}$$

the same as for the other transport processes.

For Poiseuille flow the total production of heat is

$$\begin{aligned}\frac{\mathrm{d}Q}{\mathrm{d}t} &= 2\alpha V v^2 \\ &= \frac{\mathrm{d}Q_\mathrm{S}}{\mathrm{d}t} - \frac{\mathrm{d}\Phi}{\mathrm{d}t} + V(fv\,\mathrm{d}t - fv_\mathrm{S}\,\mathrm{d}t) + V(f\,\mathrm{d}l - f\,\mathrm{d}l_\mathrm{S}),\end{aligned} \tag{8.56}$$

and in the stationary state it is $\frac{dQ_S}{dt}$. The total thermal dissipation consists of the first two terms in the second equation; the third term is the production of heat due to changes in pressure–volume work on relaxation to the stationary state, and the fourth term is the change in potential energy of the system during that relaxation.

Next we seek the stochastic equation for the probability distribution of fluctuations in the macroscopic mass velocity for which the excess work, (8.50), gives the stationary distribution. It is the Fokker–Planck equation with constant probability coefficient $c = 2\alpha kT/\rho^2$

$$\frac{\partial P(v,t)}{\partial t} = -2\frac{\alpha}{\rho}\frac{\partial}{\partial v}((v_S - v)P(v,t)) + \frac{c}{V}\frac{\partial^2}{\partial v^2}P(v,t) \tag{8.57}$$

with the stationary solution

$$P_S(v) = P_0 \exp\left(-\frac{\alpha V}{c\rho}(v_S - v)^2\right) = P_0 \exp\left(-\frac{\Phi}{kT}\right). \tag{8.58}$$

At equilibrium $v_S = 0$ and then (8.58) yields the equilibrium distribution of mass velocity.

For all the one-variable problems considered in this chapter the excess work Φ is a state function.

The generalizations to the N-block model and continuous flow for both the Couette and Poiseuille geometries proceed analogously to the cases of diffusion and thermal conduction.

8.3 Nonlinear One-Variable Transport Processes

When in the discussion of one-variable chemical kinetics we made the transition from linear to non-linear kinetic equations, Sect. 2.2, we invoked the concept of linear equivalent systems, both thermodynamically and kinetically. We shall use the same approach here and apply it only to the case of thermal conduction, as a reminder. See [1] for detailed treatments of all three transport processes.

We now pose a non-linear equation as an example

$$\frac{dT}{dt} = \kappa_1(T)(T_1 - T) - \kappa_3(T)(T - T_3), \tag{8.59}$$

where the thermal conduction coefficients κ_1 and κ_3 are functions of temperature; compare with (8.29) for a linear system where these coefficients are constants. We introduce a temperature T^*

$$T^* \equiv \frac{\kappa_1(T)T_1 + \kappa_3(T)T_3}{\kappa_1(T) + \kappa_3(T)}, \tag{8.60}$$

which we substitute for T_S in the linear problem to obtain the excess work

$$\Phi = \int^{T} C_V \left(1 - \frac{T^*}{T}\right) \mathrm{d}T, \tag{8.61}$$

which has all the properties listed in Sect. 8.1.

8.4 Coupled Transport Processes: An Approach to Thermodynamics and Fluctuations in Hydrodynamics

8.4.1 Lorenz Equations and an Interesting Experiment

We present a brief introduction to coupled transport processes described macroscopically by hydrodynamic equations, the Navier–Stokes equations [4]. These are difficult, highly non-linear coupled partial differential equations; they are frequently approximated. One such approximation consists of the Lorenz equations [5, 6], which are obtained from the Navier–Stokes equations by Fourier transform of the spatial variables in those equations, retention of first order Fourier modes and restriction to small deviations from a bifurcation of an homogeneous motionless stationary state (a conductive state) to an inhomogeneous convective state in Rayleigh–Benard convection (see the next paragraph). The Lorenz equations have been applied successfully in various fields ranging from meteorology to laser physics.

What is Rayleigh–Benard convection? Consider a fluid held between two thermally conductive parallel plates held horizontally, which is perpendicular to the gravitational field. If a temperature gradient is applied across the plates, with the bottom plate cooler, then for a given temperature difference a stable stationary homogeneous state will be attained and this is a conductive state. If, however, the lower plate is warmer than the top plate then a stationary convective state may be attained; for a given temperature difference, given fluid and given geometry of plate separation, the fluid near the top plate, being heavier than that near the bottom plate, will sink (flow) in the gravitational field. The homogeneous state has become unstable, a bifurcation occurs and a stable inhomogeneous state consisting of convective rolls becomes stable. On further increase of the temperature difference still more complicated dynamic states, including chaos, may appear.

What is the purpose of a thermodynamic and stochastic theory of hydrodynamics? The thermodynamic potential (state) functions for irreversible processes approaching equilibrium are known, for example the Gibbs free energy change for a process at constant temperature and pressure. Changes in that energy yield the maximum work, other than pressure–volume work, available from that process. Then, by analogy, the aims of a theory of thermodynamics for hydrodynamics are the establishment of evolution criteria

(Liapunov functions) with physical significance, such as the *excess* work; the work and power available from a transient decay to a stationary state; macroscopic necessary and sufficient criteria of stability of stationary states; thermodynamic criteria for bifurcations from one type of stationary state to another type; thermodynamic criteria of relative stability, that is thermodynamic criteria of state selection; and a connection of the thermodynamic theory to fluctuations.

Attempts in these directions have a long history going back to Helmholtz, Korteweg, and Rayleigh, which we shall not review here; for a comprehensive account see Lamb's classical treatise [4]. The emphasis was on the total dissipation, which, however, does not provide thermodynamic evolution criteria far from equilibrium. Glansdorff and Prigogine [4, 7] considered the second variation of the entropy as a criterion for evolution and stability. Their approach is limited to small deviations from a stationary state, provides only sufficient not necessary conditions, has no connections with work or excess work and no connections with fluctuations (master equation). Keizer formulated a stochastic approach for the relaxation to stationary states and fluctuations around a single stationary state [8]; he assumed Gaussian fluctuations, limited to small fluctuations related to linearized kinetics (for chemical kinetics). There are several articles on the statistical mechanics of stationary states and fluctuating hydrodynamics, which consists of the addition of Gaussian fluctuations to the linearized Navier–Stokes equations [9]. Here the thermodynamics may be sufficient for systems approaching equilibrium but not for stationery states far from equilibrium. In none of these studies are connections made to work or power, nor to Liapunov functions, nor to issues of relative stability when several states are available.

We developed our thermodynamic theory for the Lorenz equations, obtained with approximations from the Navier–Stokes equations (we present almost no mathematics here; that is given in detail in [10]). The Lorenz equations are

$$\begin{aligned}\dot{X} &= P(Y - X),\\ \dot{Y} &= -XZ - Y + rX,\\ \dot{Z} &= XY - bZ,\end{aligned} \tag{8.62}$$

where X represents the amplitude of the stream function of the macroscopic velocity of the fluid; Y the reduced temperature mode of the thermal conduction and Z the reduced temperature mode related to the vertical flow in the liquid layer. P is the Prandtl number (the kinematic viscosity divided by the thermal conductivity). The parameters r and b are

$$r = Rq^2/(\pi^2 + q^2)^3, \quad b = 4\pi^2/(\pi^2 + q^2), \tag{8.63}$$

where R is the Rayleigh number

$$R = \frac{\rho\alpha g h^3 \, \Delta T}{\eta\kappa} \tag{8.64}$$

and ρ is the density, α is the thermal expansion coefficient, g is the gravitational constant, h is the height of the fluid layer, ΔT is the temperature difference across the layer of liquid, η is the viscosity, and κ is the thermal conductivity. These are the variables and parameters of this system.

One solution of the Lorenz equations is $(X, Y, Z) = (0, 0, 0)$. When the control parameter r is less than unity, that is the Rayleigh number is less than its critical value R_c, then the zero solution is unique and stable, and it corresponds to the motionless conductive state of the fluid. At the bifurcation point $r = 1$ this solution becomes unstable, and a new solution becomes stable corresponding to convective modes. These solutions can be used to construct an excess work function, just as we did for single transport properties.

Now we retun to a fascinating experiment. Zamora and Ray de Luna [11] carried out Rayleigh–Benard experiments in an apparatus that could be inverted 180° in the gravitational field. In their experimental arrangement the bottom and top sides of the fluid are connected to heat reservoirs at different temperatures. Suppose a stationary state is homogeneous and conductive, achieved by setting the low temperature reservoir below the one at high temperature in the gravitational field. Next the entire apparatus is turned in the gravitational field; then the conductive state becomes unstable and a stable convective state appears. Throughout the experiment the reservoir temperatures are held fixed and the average temperature of the fluid is nearly constant. They measured the heat fluxes into and out from the reservoirs during all transients, that is the relaxations from a conductive to a convective state and the reverse. For example, say the fluid is in a convective state; invert the apparatus; the convective state is unstable and a transient change to a conductive state takes place. Measure the heat fluxes in and out of the reservoirs. The integration of the differences of the heat fluxes in and out during this transient process is the total energy change accompanying the destruction of the convective rolls. Similarly they measured the total energy change accompanying the formation of the convective rolls. The results show that the heat releases for both the destruction and formation of convective structures are *positive*, which means that the system *always* releases energy in the form of heat when it approaches a stable stationary state, either the convective state or the motionless conductive state. In an auxiliary experiment they found that the change in average temperature of the system is very small; the change in internal energy due to this small temperature is less than 10% of the heat release during the relaxation processes, and so this can not be the reason for the experimental observations. Moreover, the change in internal energy cannot explain the observed heat release in both the destruction and the formation of a convective structure.

Our theory based on the concept of exess work accounts for these experiments, at least qualitatively. According to our theory, when the system approaches a stable stationary state, either convective or conductive, there is a decrease in Φ, the excess work, and a positive excess work is released, which

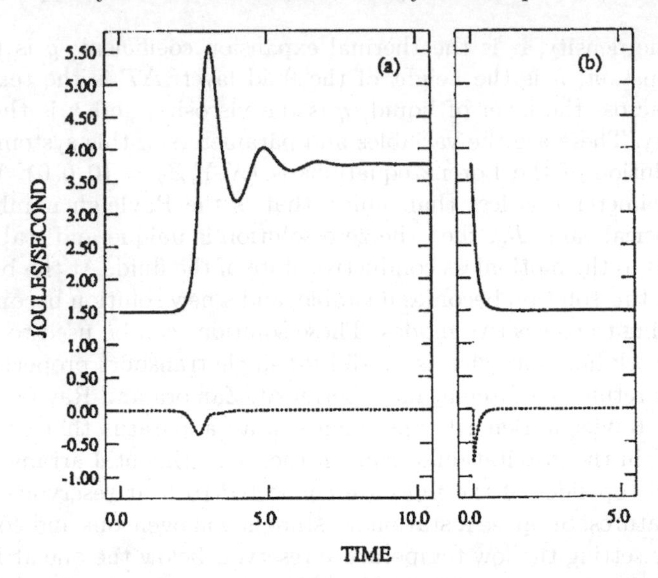

Fig. 8.5. The product of the total rate of dissipation times temperature (*solid line*) in Js^{-1} and the time derivative of excess work (*dashed line*) vs. time in the following processes for the Lorenz model: (**a**) Gravity is initially set in the direction along which the temperature decreases, and the system is at a stable motionless conductive stationary state; at $t = 0$, invert the direction of gravity; the motionless conductive state becomes unstable and the system approaches the convective stationary state. (**b**) The reverse process. The temperature difference is $|\Delta T| = 4\,\mathrm{K}$ for both cases

will be dissipated and released as heat. This is shown in a calculation from the theory in Fig. 8.5, from [10].

Figure 8.5a is for the transition from a conductive to a convective state, and Fig. 8.5b for the reverse transition. In each case the integral of $\mathrm{d}\Phi/\mathrm{d}t$ over time is negative, that is excess work is dissipated and heat is generated. Further, we note that the entropy production in the convective state is always larger than that in the conductive state, and no explanation of heat release in the transition from one state to the other can thereby be derived.

The theoretical results, based on the Lorenz model, agree with experiments qualitatively in that the total excess work change is of the same order of magnitude as the heat release measured in the experiments, and this is a major confirmation of our theory.

There may be several reasons for a lack of quantitative agreement. First, the convective stationary state in the theory is a focus, not a node. (A node is approached with an eigenvalue that is real and negative and hence provides for a damped monotonic approach, whereas a focus is approached with a complex eigenvalue with the real part negative, that is a damped oscillatory approach.) In the experiments, however, the convective stationary state is a node due to the rigid boundaries. Second, because of the truncation to the first order in Fourier modes in the Lorenz model, this model can be a good approximation near a bifurcation point from a conductive to a convective state. However, in these experiments the bifurcation point is $\Delta T = 0.95\,\mathrm{K}$ and all the data are collected between $\Delta T = 3$–$10\,\mathrm{K}$, fairly far from the bifurcation.

8.4.2 Rayleigh Scattering in a Fluid in a Temperature Gradient

In a simple fluid, in an imposed temperature gradient, light is scattered due to fluctuations in temperature and due to fluctuations in the transverse hydrodynamic velocity. Excellent experiments have been made on the measurement of this light scattering. The problem has been studied theoretically as well, by means of fluctuating hydrodynamics [9], valid for small fluctuations around a conductive state with a constant temperature gradient, which can be close to or far from equilibrium. Theory and experiment are in very good agreement [12].

In [13] we developed the relation of our studies on the thermodynamic and stochastic theory of transport properties to the reported research on this topic. There we showed that the deterministic excess work, as formulated in Chap. 2 for reactions and in this chapter for transport processes, provides a thermodynamic interpretation of fluctuations around a stationary state, either close to or far from equilibrium, for the case of Raleigh scattering from fluctuations in a fluid with an imposed temperature gradient. The stationary probability distribution is determined by a quantity proportional to the excess deterministic work. From the probability distribution we obtain, in the Gaussian approximation for small fluctuations, the matrix of correlations derived from fluctuating hydrodynamics, (8.51) in [13]. Thus in this limit of small fluctuations there is agreement between the theory of fluctuating hydrodynamics and our theory of the thermodynamics and fluctuations in transport properties.

8.5 Thermodynamic and Stochastic Theory of Electrical Circuits

This topic is mentioned but not discussed here; for its presentation see [14].

Acknowledgement. Section 8.2 of this chapter is based on ref. [1]; Sect. 8.4.1 on [9]; and Sect. 8.4.2 on [13].

References

1. J. Ross, X.-L. Chu, A. Hjelmfelt, M.G. Velarde, J. Phys. Chem. **96**, 11054 (1992)
2. L.D. Landau, E.M. Lifshitz, *Statistical Physics* (Pergamon Press, London, 1958)
3. G.K. Batchelor, *An Introduction to Fluid Dynamics* (Cambridge University Press, Cambridge, 1967)

4. H. Lamb, *Hydrodynamics*, 6th edn. (Dover Publications, New York, 1945)
5. B. Saltzman, J. Atmos. Sci. **19**, 329 (1962)
6. E.N. Lorenz, J. Atmos. Sci. **20**, 130 (1963)
7. P. Glansdorff, I. Prigogine, *Thermodynamic Theory of Structure, Stability, and Fluctuations* (Wiley, New York, 1971)
8. J. Keizer, *Statistical Thermodynamics of Nonequilibrium Processes* (Springer-Verlag, New York, 1987)
9. D. Ronis, I. Procaccia, Phys. Rev. A. **26**, 1812 (1982); R.F. Fox, Phys. Rep. **48**, 179 (1978)
10. M.G. Velarde, X.-L. Chu, J. Ross, Phys. Fluids **6**, 550 (1994)
11. M. Zamora, A. Rey de Luna, J. Fluid Mech. **167**, 427 (1986)
12. B.M. Law, J.V. Sengers, J. Stat. Phys. **57**, 531 (1987); B.M. Law, P.N. Segrè, R.W. Gammon, J.V. Sengers, Phys. Rev. A. **41**, 816 (1990); P.N. Segrè, R.W. Gammon, J.V. Sengers, B.M. Law, Phys. Rev. A. **45**, 714 (1992); W.B. Li, P.N. Segrè, R.W. Gammon, J.V. Sengers, Physica A. **204**, 399 (1994)
13. A. Suárez, J. Ross, J. Phys. Chem. **99**, 14854 (1995)
14. A. Hjelmfelt, J. Ross, Phys. Rev. A **45**, 2201 (1992)

9

Thermodynamic and Stochastic Theory for Non-Ideal Systems

9.1 Introduction

The presentations in the prior chapters have been limited to ideal systems, either gases or solutions. We now extend the thermodynamic and stochastic theory to non-ideal systems.

The formulation of the theory requires deterministic equations either of kinetics or transport for non-ideal systems. We shall illustrate the approach with the Bronsted theory of non-ideal reactants and products, which has been used extensively.

For a reaction $aA + bB$ it is assumed in the Bronsted theory [1] that the reaction proceeds through an activated complex $C^\dagger$

$$aA + bB \leftrightarrows C^\dagger \rightarrow \text{products}, \tag{9.1}$$

and that the rate of the reaction is

$$-\frac{\mathrm{d}C^\dagger}{\mathrm{d}t} = k^\dagger C^\dagger, \tag{9.2}$$

where $C^\dagger$ is the concentration of the activated complex. Furthermore, we assume (quasi)equilibrium between the reactants A and B and the activated complex

$$K^\dagger = \frac{a_{C^\dagger}}{(a_A)^a (a_B)^b} = \frac{\gamma_{C^\dagger} C^\dagger}{(\gamma_A A)^a (\gamma_B B)^b}, \tag{9.3}$$

so that the forward rate can be written

$$-\frac{\mathrm{d}C^\dagger}{\mathrm{d}t} = k^\dagger \cdot K' \frac{(a_A)^a (a_B)^b}{\gamma_{C^\dagger}} = k_f \frac{(\gamma_A)^a (\gamma_B)^b}{\gamma_{C^\dagger}} A^a B^b, \tag{9.4}$$

where a is the activity, γ is the activity coefficient and the effective rate coefficient is

$$K_\mathrm{f} = k^\dagger \cdot K^\dagger. \tag{9.5}$$

The Bronsted theory is well confirmed by experiment [2]. For electrolytes at small concentrations the activity coefficients are given by the Debye–Huckel theory. For uncharged species acitivity coefficients may be estimated by regular solution theory. Our presentation, based on the Bronsted theory, is not limited to that theory.

For a reversible chemical reaction

$$aA + bB \rightleftarrows dD + eE, \tag{9.6}$$

the rate of the reverse reaction is

$$k^{\ddagger} \cdot K^{\ddagger} \frac{(a_D)^d (a_E)^e}{\gamma_{C^{\ddagger}}} = k_r \frac{(\gamma_D)^d (\gamma_E)^e}{\gamma_{C^{\ddagger}}} D^d E^e, \tag{9.7}$$

where $C^{\ddagger}$ is the complex formed by the reverse reaction. The relation for the equilibrium constant of reaction (9.6) is

$$K = K^{\dagger}/K^{\ddagger}, \tag{9.8}$$

$$K = \frac{a_D^d a_E^e}{a_A^a a_B^b} \tag{9.9}$$

and the activated complex in the forward direction is assumed to be the same as in the reverse direction.

9.2 A Simple Example

It suffices to develop the theory for one simple example, that of a linear one-variable reaction system

$$A \underset{k_2}{\overset{k_1}{\rightleftarrows}} X \underset{k_4}{\overset{k_3}{\rightleftarrows}} B, \tag{9.10}$$

where the species A and B have either constant concentrations, as determined by measurements of concentrations, or constant activities, as determined by electrode measurements. We have a choice and we choose constant concentrations. The kinetics of the reaction for ideal systems is

$$\frac{\mathrm{d}X}{\mathrm{d}t} = k_1 A + k_4 B - (k_2 + k_3) X. \tag{9.11}$$

The system approaches either equilibrium, when $\frac{B}{A} = \frac{k_1}{k_2} \frac{k_3}{k_4}$, or a stationary state when this equality does not hold.

For non-ideal systems we use the Bronsted theory, (9.4), for each forward and reverse step of the reaction mechanism, (9.10), and obtain

$$\frac{\mathrm{d}X}{\mathrm{d}t} = \frac{k_1}{\gamma c_1} a_A + \frac{k_4}{\gamma c_4} a_B - \left(\frac{k_2}{\gamma c_2} + \frac{k_3}{\gamma c_3} \right) a_x \equiv t^+ - t^-, \tag{9.12}$$

where C_i denotes activated complexes of A, B or X. We have included the quasi-equilibrium constants of (9.3) into the respective rate coefficients. Although not essential for our development here, we assume that

$$\gamma c_1 = \gamma c_2 \quad \text{and} \quad \gamma c_3 = \gamma c_4. \tag{9.13}$$

Each activity coefficient may be a function of the concentrations of each of the species present, A, B, X, C_i. Thus the kinetic terms t^+ and t^- are non-linear functions of X, with the non-linearities due to the non-idealities.

The differential hybrid free energy for an arbitrary state with X molecules is

$$\mathrm{d}M = \mu_A \, \mathrm{d}A + \mu_X \, \mathrm{d}X + \mu_B \, \mathrm{d}B. \tag{9.14}$$

The difference between $\mathrm{d}M$ and the same quantity at the stationary state X_S for the same $A, B, \mathrm{d}A, \mathrm{d}B$, and $\mathrm{d}X$ gives the differential excess work

$$\mathrm{d}\Phi = \mathrm{d}M - \mathrm{d}M_\mathrm{S} = (\mu_x - \mu_{x_s}) \, \mathrm{d}X = RT \, \ln \frac{a_X}{a_{X_S}} \, \mathrm{d}X, \tag{9.15}$$

where the activity a_X equals the concentration X in an ideal system. For a non-ideal system we write for the chemical potentials

$$\begin{aligned}
\mu_A = \mu_A^0 + RT \, \ln a_A &= \mu_A^0 + RT \, \ln (A \cdot \gamma_A) \\
&= \mu_A|_\mathrm{ideal} + RT \, \ln \gamma_A, \\
\mu_X = \mu_X^0 + RT \, \ln a_X &= \mu_X^0 + RT \, \ln (X \cdot \gamma_X) \\
&= \mu_X|_\mathrm{ideal} + RT \, \ln \gamma_X, \\
\mu_B = \mu_B^0 + RT \, \ln a_B &= \mu_B^0 + RT \, \ln (B \cdot \gamma_B) \\
&= \mu_B|_\mathrm{ideal} + RT \, \ln \gamma_B,
\end{aligned} \tag{9.16}$$

where μ_A^0, etc. are standard chemical potentials at unit acivity. We omit the contributions of the activated complexes to the free energy because their concentrations are small.

The reaction mechanism (9.10) with non-ideal species can be mapped to a thermodynamically and kinetically equivalent ideal linear system at each instant. We require that the activities of A, X, B are the same in the two systems and the kinetic equivalency is assured by the same t^+ and t^- in the two systems, see Table 9.1, from [3].

At each value of the affinity a_X a thermodynamically and kinetically equivalent ideal linear system can be obtained from the transformations

$$\begin{aligned}
k_1^\mathrm{id} &= k_1 \frac{1}{\gamma C_1}, \quad k_2^\mathrm{id} = k_2 \frac{1}{\gamma_{C_2}}, \\
k_3^\mathrm{id} &= k_3 \frac{1}{\gamma C_3}, \quad k_4^\mathrm{id} = k_4 \frac{1}{\gamma_{C_4}},
\end{aligned} \tag{9.17}$$

Table 9.1. Mapping from a non-ideal to an ideal linear system

Ideal linear system	Nonideal linear system	
μ_A, μ_X, μ_B	μ_A, μ_X, μ_B	The same in two systems
t^+, t^-	t^+, t^-	The same in two systems
$\mathrm{d}X/\mathrm{d}t$	$\mathrm{d}X/\mathrm{d}t$	The same in two systems
a_A, a_X, a_B	a_A, a_X, a_B	The same in two systems
$A^{\mathrm{id}}, X^{\mathrm{id}}, B^{\mathrm{id}}$	A, X, B	
$k_1^{\mathrm{id}}, k_2^{\mathrm{id}}, k_3^{\mathrm{id}}, k_4^{\mathrm{id}}$	k_1, k_2, k_3, k_4	

The kinetic rate equation for the instantaneously equivalent linear and ideal system is

$$\begin{aligned}\frac{\mathrm{d}X}{\mathrm{d}t} &= k_1^{\mathrm{id}} A^{\mathrm{id}} + k_4^{\mathrm{id}} B^{\mathrm{id}} - \left(k_2^{\mathrm{id}} + k_3^{\mathrm{id}}\right) X^{\mathrm{id}} \\ &= k_1^{\mathrm{id}} a_A + k_4^{\mathrm{id}} a_B - \left(k_2^{\mathrm{id}} + k_3^{\mathrm{id}}\right) a_x = t_{\mathrm{id}}^+ - t_{\mathrm{id}}^- . \end{aligned} \tag{9.18}$$

We define the activity a_{X^*} by the relation

$$\frac{a_{X^*}}{a_X} = \frac{t^+}{t^-} . \tag{9.19}$$

This activity is a function of X and at the kinetic stationary state

$$a_{X^*}\big|_{\mathrm{st.st.}} = a_{X_{\mathrm{S}}} . \tag{9.20}$$

With the equivalence relations (9.17) we have

$$a_{X^*} = a_X \frac{t^+}{t^-} = a_X^{\mathrm{id}} \frac{t^+}{t_{\mathrm{id}}^-} = X^{\mathrm{id}} \frac{t_{\mathrm{id}}^+}{t_{\mathrm{id}}^-} = X_S^{\mathrm{id}} , \tag{9.21}$$

and we see that a_{X^*} is the activity of the stationary state of the instantaneously equivalent ideal linear system. The difference between the hybrid free energy of an arbitrary state a_X, and that of the reference state a_{X^*} for the same a_A, a_B, $\mathrm{d}A$, $\mathrm{d}B$, $\mathrm{d}X$ yields the differential excess work

$$\begin{aligned}\mathrm{d}\Phi = \mathrm{d}M - \mathrm{d}M^* &= \left(\mu_X - \mu_{X^*}\right) \mathrm{d}X \\ &= RT \ \ln \frac{a_X}{a_{X^*}} \mathrm{d}X = RT \ \ln \frac{t^-}{t^+} \mathrm{d}X , \end{aligned} \tag{9.22}$$

and the total excess work is

$$\Phi\left(X\right) = \int \left(\mu_X - \mu_{X^*}\right) \mathrm{d}X = RT \int \ln \frac{t^-}{t^+} \ \mathrm{d}X . \tag{9.23}$$

The properties of the function $\Phi(X)$ constitute the thermodynamic and stochastic theory and these properties are reviewed in Sect. 8.1, points 1–7.

The cases for more complicated chemical reaction mechanisms can be developed in a similar way without further complications, see [3].

Acknowledgement. This chapter is based on a part of ref. [3].

References

1. R.S. Berry, S.A. Rice, J. Ross, *Physical Chemistry*, 2nd edn. (Oxford University Press, New York, 2000)
2. R. Fowler, E.A. Guggenheim, *Statistical Thermodynamics* (Cambridge University Press, Cambridge, 1949)
3. J. Ross, X.-L. Chu, J. Chem. Phys. **98**, 9765 (1993)

10

Electrochemical Experiments in Systems Far from Equilibrium

10.1 Introduction

Electrochemical experiments in chemical systems at equilibrium have provided extensive information on thermodynamic quantities, such as Gibbs free energy changes and related equilibrium constants [1]. At equilibrium the Gibbs free energy change of a chemical reaction is related to the voltage (the electrochemical potential) generated by that reaction run in an electrochemical cell

$$\Delta G = -nFE, \tag{10.1}$$

where ΔG is the Gibbs free change of the reaction, n is the number of equivalents of electrons transferred from one electrode to the other for the stoichiometric reaction as written and E is the electrochemical potential.

In Chap. 2–9 we presented a thermodynamic and stochastic theory of chemical reactions and transport processes in non-equilibrium stationary and transient states approaching non-equilibrium stationary states. We established a state function Φ, which for systems approaching equilibrium reduces to ΔG. Since Gibbs free energy changes can be determined by macroscopic electrochemical measurements, we seek a parallel development for the determination of Φ by macroscopic electrochemical and other measurements.

We begin with a discussion of two kinds of experiments and then, in the next chapter, turn to the development of the thermodynamic and stochastic theory to connect with these experiments. There are very few thermodynamic measurements on complex open reaction systems (see [2] as one example).

10.2 Measurement of Electrochemical Potentials in Non-Equilibrium Stationary States

When chemical species come into equilibrium with an electrode in an open circuit, the potential between the electrode and a reference electrode is related to the potential difference of the half reaction occurring at the electrode. If no other reactions are occurring then this potential is related to the Gibbs free energy difference of the half reaction at the electrode. If there are other reactions occurring then the species may be in non-equilibrium states, even though they are in equilibrium with the electrode, and the potential is that of a non-equilibrium stationary state. If local equilibrium holds then the potential is the Gibbs free energy difference; if it does not hold, in that there are degrees of freedom, such as the reactions, which are explicitly held away from equilibrium, then deviations from the Gibbs free energy difference may occur. We shall speak of Nernstian, (10.1), and non-Nernstian contributions to the electrochemical potential. There is one prior measurement of the type to be described and that is by Keizer and Chang [3], following a suggestion by Keizer [4] that there should be a non-Nernstian contribution to the electrochemical potential in nonlinear reaction systems approaching to, or in, stationary states far from equilibrium. They reported a very small non-Nernstian contribution in an Fe(II)/Fe(III) reaction system in a non-equilibrium stationary state.

We studied the autocatalytic minimal bromate reaction, which can be oscillatory, but was studied in a bistable regime. A proposed mechanism for this reaction, and participating species, are listed in Table 10.1.

The net reaction is the oxidation of Ce(III) to Ce(IV) by bromate. In the bistable regime there is a state, where essentially no reaction occurs, which coexists with a state in which a percentage of Ce(III) is oxidized to Ce(IV). In this system we measured [6] at the same time the optical density which gives concentrations of Ce(IV) by Beer's law, and hence also the concentration of Ce(III) by conservation, and the emf of a Pt electrode which at equilibrium follows the Nernst equation (10.1). The experiment consisted of the measurement of the emf of the Ce(III)/Ce(IV) half reaction at a redox (Pt-Ag/AgCl) electrode under equilibrium and stationary non-equilibrium conditions. The apparatus is shown in Fig. 10.1, but in these experiments the parts 4–7 were not present. From these measurements we determined that there exists a non-Nernstian contribution in a non-equilibrium stationary state as shown in Table 10.2. The concentration of $[Ce(III)]_{ss}$ in the stationary state is obtained

Table 10.1. Reaction mechanism for the minimal bromate reaction, from (5)

Number	Reaction
(B1)	$BrO_3^- + Br^- + 2H^+ \rightleftarrows HBrO_2 + HOBr$
(B2)	$HBrO_2 + Br^- + H^+ \rightleftarrows 2HOBr$
(B3)	$HOBr + Br^- + H^+ \rightleftarrows Br_2 + H_2O$
(B4)	$BrO_3^- + HBrO_2 + H^+ \rightleftarrows 2BrO_2^{\cdot} + H_2O$
(B5)	$Ce^{3+} + BrO_2^{\cdot} + H^+ \rightleftarrows Ce^{4+} + HBrO_2$
(B6)	$Ce^{4+} + BrO_2^{\cdot} + H_2O \rightleftarrows Ce^{3+} + BrO_3^- + 2H^+$
(B7)	$2HBrO_2 = BrO_3^- + HOBr + H^+$

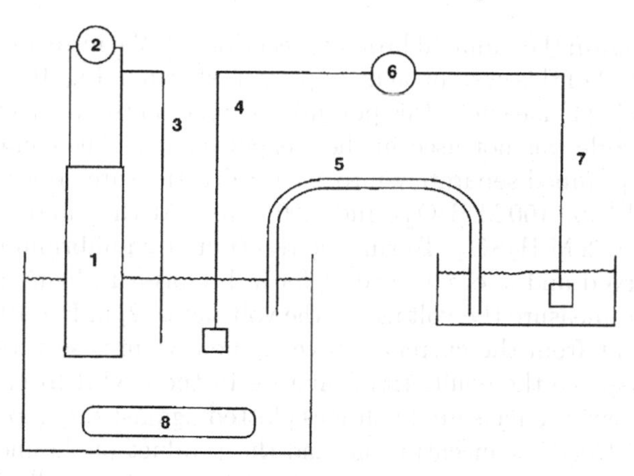

Fig. 10.1. Schematic drawing of apparatus: 1, combination electrode Pt, Ag–AgCl; 2, voltmeter; 3, Br^- electrode; 4, Pt electrode; 5, salt bridge; 6, constant current source; 7, Pt electrode; 8, stirrer bar. The inlet flow and aspiration outlet on the CSTR (left vessel) are not shown. CSTR – continuous stirred tank reactor. From (7)

Table 10.2. Results from the minimal bromate experiment with various concentrations of Ce(III) in the combined feedstreams into the reactor, $[\mathrm{Ce(III)}]_0$, from [6]

$[\mathrm{Ce(III)}]_0$ (M)	$[\mathrm{Ce(III)}]_{ss}$ (M)	IE emf (mV)	M emf (mV)
1.500×10^{-3}	1.393×10^{-3}	1,167.0	1,180.2
1.397×10^{-3}	1.361×10^{-3}	1,176.0	1,183.0
1.297×10^{-3}	1.277×10^{-3}	1,187.5	1,189.7
8.333×10^{-4}	8.700×10^{-4}	1,223.7	1,223.8

from the absorption measurement. The local equilibrium emf, the third column in Table 10.2 is calculated from the ratio Ce(III)/Ce(IV) and the Nernst equation. The measured emf in the fourth column of the table is that measured by the Pt electrode. The difference is small, about 1% of the emf at the largest inflow concentration of $\mathrm{Ce(III)}_0$ and decreases for smaller inflow concentrations.

10.3 Kinetic and Thermodynamic Information Derived from Electrochemical Measurements

In the experiment to be described we study the electrochemical displacement of a non-linear chemical system, the minimal bromate reaction, from non-equilibrium stationary states and from equilibrium. In the following chapter we shall relate such measurements to the thermodynamic and stochastic theory of potentials governing fluctuations in electrochemical systems in stationary states far from, near to and at equilibrium.

We study again the minimal bromate reaction [7]. We measure the Ce(III)/Ce(IV) potential of this system on an apparatus shown in Fig. 10.1. The combination electrode (1) measures this potential as read on the voltmeter (2). (The bromide electrode was not used in these experiments.) The contents of three reservoirs are pumped separately into the CSTR; the three reservoirs contain 0.00450 M Ce^{III}, 0.0100 M BrO_2^-, and 1.00×10^{-6} M Br^-, and each reservoir contains also 0.72 M H_2SO_4. To run the reaction at equilibrium the three solutions are mixed and allowed to react for a day prior to being pumped into the CSTR. We measure the voltage on the voltmeter, 2, in Fig. 10.1, with zero imposed current from the current source 6; then we impose various currents with 6 and displace the equilibrium mixture in the CSTR from equilibrium. Figure 10.2 shows the measured voltages plotted against the imposed current, as well as the Ce(IV) concentration, and the product of the measured voltage minus the stationary state voltage, in this case the equilibrium voltage, multiplied by the imposed current.

A non-equilibrium stationary state is achieved by flowing the reacting solutions into the CSTR at given flow rates, that is given residence times in the reactor; the measurements just described are repeated, and shown for a residence time of 175 s in Fig. 10.3, and a residence time of 400 s in Fig. 10.4.

First, it is interesting to compare the equilibrium displacement plot (Fig. 10.2) with the plots of displacements from non-equilibrium stationary

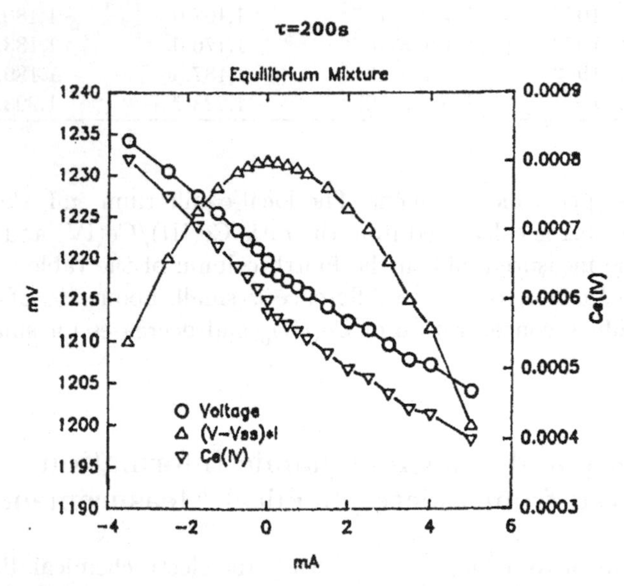

Fig. 10.2. Plot of voltage V, as measured on 2 in Fig. 10.1, the power input $(V - V_{ss})I$ and the Ce(IV) concentration vs. the imposed current I. V_{ss} is the measured voltage at the stationary state, here equilibrium, at zero imposed current. The residence time in the CSTR is 200 s, from [7]

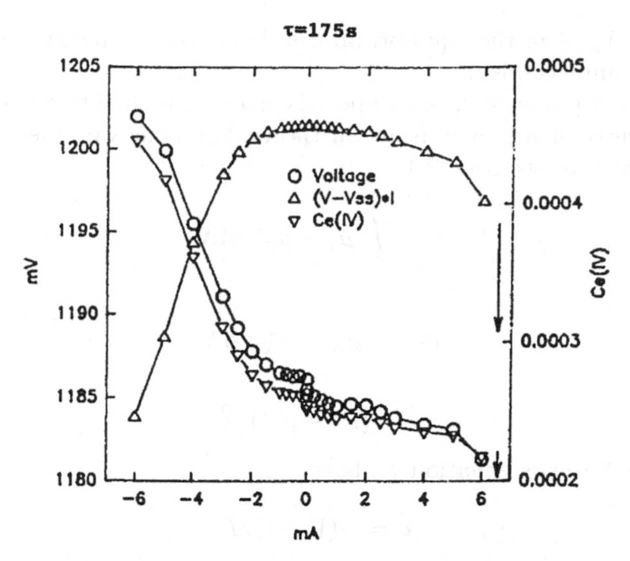

Fig. 10.3. Experiments as in Fig. 10.2 for a non-equilibrium stationary state at zero imposed current and displacements from that state with imposed currents. The residence time is 175 s. The arrows indicate transitions to other stationary states, from [7]

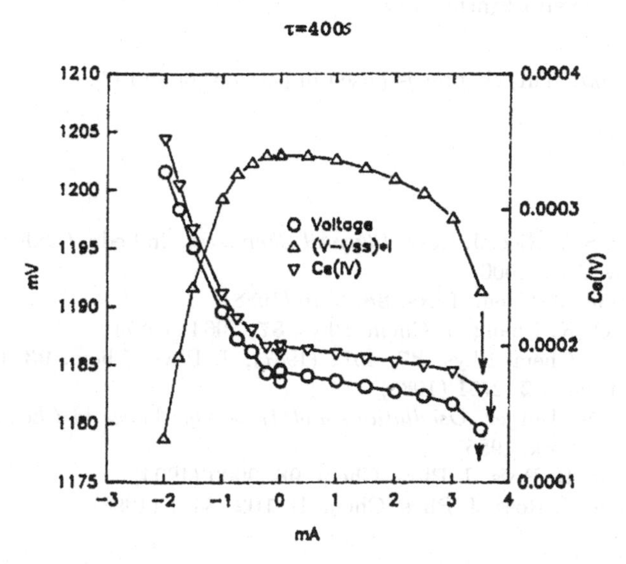

Fig. 10.4. Same as Fig. 10.3, but the residence time is 400 s, from [7]

states (Figs. 10.3 and 10.4). To achieve a given displacement, either in Ce(IV) concentration or in the potential from its stationary value at zero current, a larger imposed current is necessary in the non-equilibrium cases, Figs. 10.3 and 10.4, than in the equilibrium case, Fig. 10.2. We further note that the plot of $(V - V_{ss})I$ in the equilibrium case is nearly symmetric, those on the non-equilibrium cases is not.

The plots of power input vs. imposed current can be obtained by a simple, nearly dimensional argument from our theory. For a one-variable linear system the excess work is, see Sect. 2.1,

$$\phi = \int (\mu_x - \mu_x{}^{\mathrm{s}})\, \mathrm{d}X \tag{10.2}$$

hence $\mathrm{d}\phi$ is

$$\mathrm{d}\phi = (\mu_X - \mu_X^s)\, \mathrm{d}X \tag{10.3}$$

and

$$\dot{\phi} = (\mu_x - \mu_x{}^{\mathrm{s}})\, \dot{X} \tag{10.4}$$

so that with Nernst's equation we have

$$\dot{\phi} = -(V - V_{\mathrm{s}})I \tag{10.5}$$

The time derivative of the excess work, which is that part of the dissipation due to the variation is X, equals the power input necessary to maintain the system away from its stationary state. In the next chapter we develop the theory in a more substantial way.

Acknowledgement. This chapter is based in part on [6] and [7].

References

1. R.S. Berry, S.A. Rice, J. Ross, *Physical Chemistry*, 2nd edn. (Oxford University Press, New York, 2000)
2. M.G. Roelofs, J. Chem. Phys. **88**, 5516 (1988)
3. J. Keizer, O.-K. Chang, J. Chem. Phys. **87**, 4064 (1987)
4. J. Keizer, J. Chem. Phys. **87**, 4074 (1987); J. Phys. Chem. **93**, 6939 (1989); J. Chem. Phys. **82**, 2751 (1985)
5. R.J. Field, M. Burger, *Oscillations and Traveling Waves in Chemical Systems* (Wiley, New York, 1985)
6. A. Hjelmfelt, R. Ross, J. Phys. Chem. **98**, 9900 (1994)
7. A. Hjelmfelt, J. Ross, J. Phys. Chem. B. **102**, 3441 (1998)

11

Theory of Determination of Thermodynamic and Stochastic Potentials from Macroscopic Measurements

11.1 Introduction

We have several purposes in mind for this chapter: First we present a development of the theory for the determination of thermodynamic and stochastic potentials, ϕ, for non-equilibrium systems from electrochemical measurements; second, a parallel development for neutral (not ionic) systems in general; and third, the presentation of suggestions for testing the consistency of the master equation with such measurements [1].

For the first purpose we choose a chemical reaction system with some ionic species, as for example the minimal bromate reaction, for which we presented some experiments in Chap. 10. The system may be in equilibrium or in a non-equilibrium stationary state. An ion selective electrode is inserted into the chemical system and connected to a reference electrode. The imposition of a current flow through the electrode connection drives the chemical system (CS) away from its initial stationary state to a new stationary state of the combined chemical and electrochemical system (CCECS), analogous to driving the CS away from equilibrium in the same manner. A potential difference is generated by the imposed current, which consists of a Nernstian term dependent on concentrations only, and a non-Nernstian term dependent on the kinetics. We shall relate the potential difference to the stochastic potential; for this we need to know the ionic species present and their concentrations, but we do not need to know the reaction mechanism of the chemical system, nor rate coefficients.

For the second purpose we offer a suggestion for reaction systems with or without ionic species for an indirect method of determining the stochastic potential from macroscopic measurements. We impose an influx of any of the stable intermediate chemical species into the system (CS), and thus displace the CS from its initial stationary state to a new stationary state of the combined CS and the imposed influx. We measure the concentrations of species in the new stationary state and repeat this experiment for different imposed influx rates. We can fit these measurements to an assumed reaction mechanism and an assumed master equation to determine rate coefficients and the stochastic potential.

11.2 Change of Chemical System into Coupled Chemical and Electrochemical System

Let us consider a simple model system

$$R_+ + Q_+ \underset{k_{+1}^-}{\overset{k_{+1}^+}{\rightleftarrows}} A_1 + B_2 \underset{k_0^-}{\overset{k_0^+}{\rightleftarrows}} B_1 + A_2 \underset{k_{-1}^-}{\overset{k_{+1}^+}{\rightleftarrows}} R_- + Q_- \tag{11.1}$$

where R_+, Q_+, R_-, and Q_- are held at given constant concentrations R_+, Q_+, R_-, Q_- respectively. The species A_1, A_2, B_1 and B_2 (at concentrations A_1, A_2, B_1, and B_2) respectively, are uniformly distributed in a CSTR. The deterministic kinetic equations are

$$\begin{aligned}\frac{\mathrm{d}A_1}{\mathrm{d}t} &= \frac{\mathrm{d}B_2}{\mathrm{d}t} = k_{+1}^+ R_+ Q_+ + k_0^- B_1 A_2 - (k_{+1}^- + k_0^+) A_1 B_2 \\ \frac{\mathrm{d}A_2}{\mathrm{d}t} &= \frac{\mathrm{d}B_1}{\mathrm{d}t} = k_{-1}^- R_- Q_- + k_0^+ A_1 B_2 - (k_{-1}^+ + k_0^-) B_1 A_2\end{aligned} \tag{11.2}$$

and for values of R_+Q_+/R_-Q_- not equal to the equilibrium constant for the overall reaction a non-equilibrium stationary state is achieved in time. The concentrations not given are assumed to be measurable.

To move this chemical system (CS) away from its stationary state we insert an electrode into the CS and connect this electrode to a reference electrode, see Fig. 10.1 in Chap. 10. The electrode reactions are

$$A_1 + e^- \rightleftarrows A_2, \quad B_1 + e^- \rightleftarrows B_2 \tag{11.3}$$

We consider a combination of the electrode reactions in (11.3) with the buffered species R_+, Q_+, R_-, Q_-,

$$R_+ \rightleftarrows A_1 + e^- \rightleftarrows A_2 \rightleftarrows R_-; \; Q_+ \rightleftarrows B_1 + e^- \rightleftarrows B_2 \rightleftarrows Q_-. \tag{11.4}$$

The sum of the reactions in (11.3) and (11.4) is the chemical reaction in (11.2). Upon insertion of the electrode into CS and the imposition of a current flow through the external circuit a new stationary state is achieved in time, a stationary state of the CCECS.

The expression for a given imposed current j is given by the Butler–Volmer equations, (11.2) and (11.3). These are essentially empirical equations [2, 3]; Bard and Faulkner refer to them as the Butler–Volmer 'approach'. According to those equations the imposed current is given by

$$j = j_1^- - j_1^+ + j_2^- - j_2^+ + j_3^- - j_3^+ + j_4^- - j_4^+ + j_5^- - j_5^+ + j_6^- - j_6^+, \tag{11.5}$$

where

$$j_1^+ = \mathscr{F} C_{R_+A_1} \exp[-\Delta G^{\ddagger}_{R_+A_1}(0)/RT] \exp(-\alpha_{R_+A_1} fE)$$
$$j_1^- = \mathscr{F} C_{A_1R_+A_1} \exp[-\Delta G^{\ddagger}_{R_+A_1}(0)/RT] \exp[(1-\alpha_{R_+A_1}) fE], \text{ etc.} \quad (11.6)$$

In these equations E is the potential difference between the two electrodes, F is the Faraday constant; C_{R+} etc. are Arrhenius prefactors; R_+, A_1 are concentrations; $\Delta G^{\ddagger}_{R_+A_1}(0)$ is the Gibbs free energy of activation of the electrode reaction $R_+ = A_1 + e^-$ when the potential difference between the electrode and the reacting solution vanishes, and similarly for the other activation energies; $a_R + A_1$ is the transfer coefficient for the R_+/A_1 reaction etc., and $f = F/RT$. In (11.5) the first term is the current due to the half-cell reaction

$$R_+ \to A_1 + e^- \quad (11.7)$$

and the second term due to

$$A_1 + e^- \to R_+ \quad (11.8)$$

and similarly for the other reactions in (11.5).

The chemical kinetics of the CCECS are given by a combination of the chemical terms in (11.2) and the electrochemical terms in (11.5)

$$\frac{dA_1}{dt} = k_{+1}^+ R_+ Q_+ + k_0^- B_1 A_2 - (k_{+1}^- + k_0^+) A_1 B_2 + \frac{1}{\mathscr{F}}(j_1^+ - j_1^- + j_2^- - j_2^+),$$
$$\frac{dA_2}{dt} = k_{-1}^- R_- Q_- + k_0^+ A_1 B_2 - (k_{-1}^+ + k_0^-) B_1 A_2 + \frac{1}{\mathscr{F}}(j_2^+ - j_2^- + j_3^- - j_3^+),$$
$$\frac{dB_1}{dt} = k_{-1}^- R_- Q_- + k_0^+ A_1 B_2 - (k_{-1}^+ + k_0^-) B_1 A_2 + \frac{1}{\mathscr{F}}(j_4^+ - j_4^- + j_5^- - j_5^+),$$
$$\frac{dB_2}{dt} = k_{+1}^+ R_+ Q_+ + k_0^- B_1 A_2 - (k_{+1}^- + k_0^+) A_1 B_2 + \frac{1}{\mathscr{F}}(j_6^- - j_6^+ + j_5^+ - j_5^-).$$
$$(11.9)$$

For a stationary state of the CCECS we have

$$\frac{dA_1}{dt} = \frac{dA_2}{dt} = \frac{dB_1}{dt} = \frac{dB_2}{dt} = 0. \quad (11.10)$$

The concentrations of the chemical species: A_1, A_2, B_1, B_2 in the stationary state of the CCECS are altered from those in the CS due to imposition of a current. Thus by varying the input current j, stationary (without electrodes) and non-stationary (with electrodes), states of the CS are all time independent and hence the concentrations are easy to measure.

11.3 Determination of the Stochastic Potential ϕ in Coupled Chemical and Electrochemical Systems

The differential of the stochastic potential for the chemical system in one of its stationary states, $d\phi_c$, is (see Chap. 3)

$$d\Phi_c = \sum_i (\mu_i - \mu_i^0) dn_i \tag{11.11}$$

with the index i extending over A_1, B_1, taken to be neutral, and the species A_2, B_2 taken to be negatively charged. The chemical potentials μ_i^0 are those of the species in a reference state; the concentrations x_j of the species X_j is obtained from the momentum canonically conjugate to x_j along a fluctuational trajectory, see Chap. 3. The differential $d\Phi_c$ is exact and any path of integration suffices to obtain Φ. The exponential of the integral in (11.11) is a formal representation of the eikonal approximation for the stationary solution of the master equation of the chemical system.

We choose the same variables for the chemical as for the electrochemical system. The reactions at the electrodes are sufficiently fast that the measured potential is the equilibrium potential; fluctuations in that potential and in the imposed current are neglected. This is analogous to neglecting fluctuations in concentrations of species in equilibrium with mass reservoirs. For systems for which equilibrium is the only stable attractor, the chemical potential of each chemical species, say that of A_2, is

$$\mu_{A_2} + E_{A_2}\mathscr{N}\mathscr{F}, \tag{11.12}$$

where E_{A_2} is the potential for a given imposed current, N is the number of equivalents in the half-cell reaction for A_2, and F is the Faraday constant. We postulate that we may write $d\Phi_E$ for the CCECS in a parallel way

$$\begin{aligned} d\Phi_E &= (\mu_{A_1} - \mu_{A_1}^0) dn_{A_1} + (\mu_{B_1} - \mu_{B_1}^0) dn_{B_1} \\ &\quad + (\mu_{A_2} + E_{A_2}\mathscr{N}\mathscr{F} - \mu_{A_2}^0 - E_{A_2}^0\mathscr{N}\mathscr{F}) dn_{A_2} \\ &\quad + (\mu_{B_2} + E_{B_2}\mathscr{N}\mathscr{F} - \mu_{B_2}^0 - E_{B_2}^0\mathscr{N}\mathscr{F}) dn_{B_2} \\ &= d\Phi c + (E_{A_2} - E_{A_2}^0)\mathscr{N}\mathscr{F} dn_{A_2} + (E_{B_2} - E_{B_2}^0)\mathscr{N}\mathscr{F} dn_{B_2}, \end{aligned} \tag{11.13}$$

where the first line is for the neutral species and the next two line for the ionic species. The fourth line gives the relation of the stochastic potential of the combined systems to that of the chemical system.

This postulate is consistent to given approximations with an expansion of the master equation or an equivalent Hamilton–Jacobi equation. The derivation is given in Appendix A and B of [1]; the mathematics is complex and specialized. The study of stochastic equations of electrochemical systems is in its infancy; we know of no prior work on this subject. Further intensive studies will be necessary to fully substantiate the postulate of (11.13).

At a stationary state of the combined system the differential $d\Phi_E = 0$ and therefore we have for $d\Phi_c$ the result

$$\begin{aligned} d\Phi_c = &- \mathcal{N}\mathcal{F}[(E_{A_2} - E_{A_2(s)}) - (E^0_{A_2} - E_{A_2(s)})]dn_{A_2} \\ &- \mathcal{N}\mathcal{F}[(E_{B_2} - E_{B_2(s)}) - (E_{B_2} - E_{B_2(s)})]dn_{B_2}, \end{aligned} \quad (11.14)$$

where we added and subtracted the potentials $E_{A_2(s)}$ of A and $E_{B_2(s)}$ of B in the stationary state of the system of the chemical system. The first term in each square bracket depends on concentrations only and thus is the Nernstian contribution to the measured electrochemical potential. The second term in each square bracket depends on the kinetics of the chemical system and thus is the non-Nernstian contribution the electrochemical potential. Measurements of this potential, say with ion-specific electrodes, yield the slopes $\partial\Phi_c/\partial n_{A_2}$ and $\partial\Phi_c/\partial n_{B_2}$; thus with measurements of the macroscopic concentrations of A_1, A_2, B_1 and B_2 at a sufficient number of displacements from the stationary state of the chemical system we can determine the stochastic potential of that system from macroscopic measurements. (The experimental results reported in Chap. 10 are insufficient for this purpose. More experimentation is in progress in the laboratory of Prof. K. Showalter at the University of West Virginia and at its completion a comparison of the stochastic potential as given by the theory and the experiments can be made.) To obtain these results no direct use of any master equation has been made and no model of the reaction mechanism was necessary.

11.4 Determination of the Stochastic Potential in Chemical Systems with Imposed Fluxes

Consider the chemical system in (11.1) with the species being either ions or neutrals; the system is in a reaction chamber in a non-equilibrium stationary state. We impose a flux of species A_1, $J = k'A'_1$, into the reaction chamber with Q_+ and Q_- held constant and thereby move the chemical system to a different non-equilibrium stationary state with different concentrations of the reacting species A_1, B_1, A_2, B_2. This procedure allows the sampling of different combinations of the reacting species by means of the imposition of different fluxes of these reactants. These combinations represent different non-stationary states in the absence of imposed fluxes, but with the imposed fluxes they are stationary states and hence measurements may be made without constraints of time. If we would attempt to measure concentrations in non-stationary states then the measurement technique would have to be fast compared to the time scale of change of the concentrations due to chemical reactions.

Now we impose a flux of a chemical species present in the system and inquire on the effect of that imposition on the stochastic potential of the system. For that we need to go from the deterministic kinetic equations to a stochastic equation, say the lowest order eikonal approximation to the chemical master equation. The response measurements to the imposed flux provide an indirect determination of the stochastic potential, one that depends on the use of the master equation, an assumed reaction mechanism, and assumed rate coefficients.

This procedure is easy for a one-variable system because we know the solution of the stationary master equation to this approximation. For example, for the one-variable Schlögl model we have the elementary reaction steps

$$A + 2X \underset{k_2}{\overset{k_1}{\rightleftarrows}} 3X, \quad X \underset{k_4}{\overset{k_3}{\rightleftarrows}} B, \tag{11.15}$$

with the concentrations of A and B held constant. The kinetic equation without the imposed flux is

$$\frac{\mathrm{d}X}{\mathrm{d}t} = k_1 A X^2 + k_4 B - (k_2 X^3 + k_3 X) \equiv t^+ - t^- . \tag{11.16}$$

Let the imposed flux be $J = k'X'$. The stationary solution of the lowest order eikonal approximation of the master equation for the system (11.15) with the imposed flux is

$$P(X) \approx \exp\left(-\frac{\Phi'}{k_B T}\right), \tag{11.17}$$

where

$$-\frac{1}{k_B T V}\frac{\partial \Phi'}{\partial X} = \ln\left(\frac{t^+ + J}{t^-}\right) = \ln\left(\frac{t^+}{t^-}\right) + \ln\left(1 + \frac{J}{t^-}\right). \tag{11.18}$$

In the absence of an imposed flux the solution reduces to

$$-\frac{1}{k_B T V}\frac{\partial \Phi}{\partial X} = \ln\left(\frac{t^+}{t^-}\right). \tag{11.19}$$

At a stationary state of the system with imposed flux we have $\partial\Phi'/\partial X = 0$ and hence from (11.18) we obtain

$$-\frac{1}{k_B T V}\frac{\partial \Phi}{\partial X} = \ln\left(\frac{t^+}{t^-}\right) = -\ln\left(1 + \frac{J}{t^-}\right). \tag{11.20}$$

Thus from measurements with imposed flux we obtain the derivative of the stochastic potential for the system without imposed flux, but we need kinetic information, the rate coefficients in t^-, as well.

For multi-variable systems this approach is more difficult; the determination of the stochastic potential requires sufficient measurements to determine rate coefficients and then the numerical solution of the stationary form of the master equation. Details of this procedure are described in Appendix A of [1].

11.5 Suggestions for Experimental Tests of the Master Equation

A direct test of the master equation for systems in non-equilibrium stationary states comes from the measurements of concentration fluctuations; such measurements have not been made yet. Some other tests of the master equation are possible based on the earlier sections in this chapter, where we can compare measurements of the stochastic potential with numerical solutions of the master equation (which requires knowledge of rate coefficients and the reaction mechanism of the system).

There are other indirect methods. Consider a one-variable system (or an effectively one variable). Let the system have multiple stationary states and in Fig. 11.1, taken from [1], we show a schematic diagram of the hysteresis loop in such systems.

Several experiments can be suggested to test aspects of the predictions of the master equation. To construct a diagram as in Fig. 11.1 from the master equation we need to know or guess rate coefficients and the reaction mechanism of the system. For the experiments we need to measure the concentration c of a given species as the the influx coefficient is varied. Thus we establish the solid lines by experiment. If we can form a CCECS, as discussed in Sects. 11.2 and 11.3 of this chapter, then we can locate the combined system at point 1 on line A by imposing a given current flow. This point is a stable stationary state of the combined system. If the imposed current is stopped (the electrochemical system is disconnected) then the chemical system will return deterministically

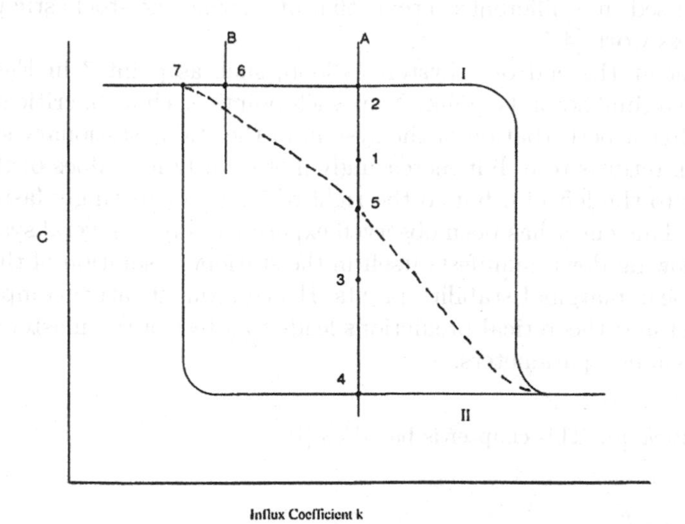

Fig. 11.1. Typical hysteresis loop for a one-variable system with a cubic kinetic equation: plot of concentration c vs. influx coefficient. *Solid lines*, stable stationary states (nodes); *broken line*, unstable stationary state. For a discussion of lines A and B and numbers, see the text

to the nearest stable stationary state of the chemical system, that is point 2 on line A and on the stable branch I. We can repeat this experiment by locating the CCECS, say on point 3 on line A; then on stopping the imposed current the chemical system will return to point 4 on line A and on the stable branch II. By means of such experiments we can locate the branch of unstable stationary states, the separatrix, the dotted line in Fig. 11.1, and compare it with predictions of the master equation.

The same approach works for the displacement of a system by imposition of an influx of a given species, see Sect. 11.4 of this chapter.

Equistability of a homogeneous stable stationary state on the upper branch of the hysteresis loop, labelled I in Fig. 11.1, with a homogeneous stable stationary state on the lower branch, labelled II, occurs at one value of the influx coefficient k within the loop. Say that point occurs at the location of line A. The predictions of the stationary solution of the master stochastic master equation are (a) the minimum of the bimodal stationary probability distribution is located on the separatrix, and (b) at equistability the probability of fluctuations $P(c)$ obeys the condition

$$\int_4^5 P(c)\mathrm{d}c = \int_2^5 P(c)\mathrm{d}c. \tag{11.21}$$

Approximately, at equistability the height of the probability peak at point 2 equals that at point 4. To either side of the value of k at equistability, the peak of the more stable stationary state is higher than the other peak. A comparison of deterministic and stochastic calculations (not experiments) has been discussed in a different context, that of viewing the stochastic potential as an excess work [4,5].

A point at the end of a hysteresis loop, such as point 7 in Fig. 11.1, is called a marginal stability point. Near such points, such as 6, critical slowing occurs. After a perturbation of the system in the stable stationary state at 6 the system returns to 6, but increasingly more slowly for values of the influx coefficient to the left of 6 but to the right of 7, and increasingly faster to the right of 6. This effect has been observed experimentally in several systems [6]. Critical slowing down manifests itself in the stationary solution of the master equation near marginal stability points. Hence a quantitative comparison of experiment and theoretical predictions leads to a test of the master equation and the assumed parameters.

Acknowledgement. This chapter is based on [1].

References

1. J. Ross, K.L.C. Hunt, M.O. Vlad, J. Phys. Chem. A **106**, 10951–10960 (2002)
2. A.J. Bard, L.R. Faulkner, *Electrochemical Methods: Fundamentals and Applications*, 2nd edn. (Wiley, New York, 2001)

3. R.S. Berry, S.A. Rice, J. Ross, *Physical Chemistry*, 2nd edn. (Oxford University Press, Oxford, 2000) Chapter 31, 13
4. B. Peng, K.L.C. Hunt, P.M. Hunt, A. Suarez, J. Ross, J. Chem. Phys. **102**, 4548–4562 (1995)
5. J. Ross, K.L.C. Hunt, P. Hunt, J. Chem. Phys. **96**, 618–629 (2002)
6. J. Kramer, J. Ross, J. Chem. Phys. **83**, 6234–6241 (1985)

Part II

Dissipation and Efficiency in Autonomous and Externally Forced Reactions, Including Several Biochemical Systems

12

Dissipation in Irreversible Processes

12.1 Introduction

We have introduced the concept of dissipation in Chap. 2.3; it is related to the entropy production in irreversible chemical and physical systems, and has been discussed in the scientific literature and in texts extensively. Here we present some relatively new developments and return to the issue of dissipation and efficiency in chemical reactions in Chap. 13 and later.

We begin with several exact solutions for the entropy production in simple irreversible processes [1].

12.2 Exact Solution for Thermal Conduction

12.2.1 Newton's Law of Cooling

We consider a macroscopic, homogeneous system with cylindrical shape of length l and cross-sectional area A; the temperature of the cylinder is T, which is generally time-dependent. Heat is transported across the boundary of the cylinder at each end in contact with two thermal baths, one with temperature T_1 at one end, and the other with T_2 at the other end; without restrictions we assume $T_1 > T_2$. We take the conduction of heat to be given by Newton's equations

$$lA\rho c \frac{\mathrm{d}T}{\mathrm{d}t} = k\left(\frac{T_1 - T}{l}\right) + k\left(\frac{T_2 - T}{l}\right), \tag{12.1}$$

where l is the length of the cylinder, A the cross-sectional area, k the thermal conductivity, ρ the density of the system taken to be constant and c the mass specific heat capacity, also taken to be constant. The solution of this equation is

$$T(t) = T_{\mathrm{st}} + [T_0 - T_{\mathrm{st}}]\exp[-2 \in t], \tag{12.2}$$

where the relaxation rate is

$$2 \in = 2k/(\rho c A l^2) \tag{12.3}$$

and the stationary state temperature in the limit of long time is

$$\lim_{t\to\infty} T = T_{\rm st} = (T_1 + T_2)/2. \tag{12.4}$$

The rate of entropy production is the product of the heat flux times the conjugated force

$$\begin{aligned}\sigma &= Alk.\frac{T_1 - T}{l}\cdot\frac{1}{l}\left(\frac{1}{T}-\frac{1}{T_1}\right) + Alk.\frac{T_2 - T}{l}\cdot\frac{1}{l}\left(\frac{1}{T}-\frac{1}{T_2}\right)\\ &= \frac{kA}{TT_1T_2l}[T^2(T_1+T_2) - 4T_1T_2T + T_1T_2(T_1+T_2)] \geq 0\end{aligned} \tag{12.5}$$

and is always positive. Note that the flux is not proportional to the force.

The derivative of the entropy production with respect to temperature is

$$\frac{\mathrm{d}}{\mathrm{d}T}\sigma(T) = \frac{Ak(T^2 - T_1T_2)(T_1+T_2)}{lT^2T_1T_2} \tag{12.6}$$

and at the stationary state we have

$$\sigma(T_{\rm st}) = Ak(T_1 - T_2)^2/2lT_1T_2. \tag{12.7}$$

Hence we may rewrite (12.5) as

$$\frac{\mathrm{d}}{\mathrm{d}T}\sigma(T_{\rm st}) = \frac{Ak(T_1-T_2)^2}{lT_1T_2(T_1+T_2)} \tag{12.8}$$

which is always positive and therefore the rate of entropy production is never an extremum at a stationary state, whether close to or far from equilibrium, except at equilibrium. This will be discussed further, later in this chapter.

At equilibrium, at $T_1 = T_2$, the entropy production rate has an extremum which is a minimum.

The dimensionless ratio

$$\left[\frac{\sigma(T)/T}{\mathrm{d}\sigma(T)/\mathrm{d}T}\right]_{T=T_{\rm st}} = 1 > 0 \tag{12.9}$$

is positive for any non-equilibrium stationary state; at equilibrium this ratio is zero

$$\left[\frac{\sigma(T)/T}{\mathrm{d}\sigma(T)/\mathrm{d}T}\right]_{T=T_{\rm eq}} = \left[\frac{T-T_{\rm eq}}{T+T_{\rm eq}}\right]_{T=T_{\rm eq}} = 0. \tag{12.10}$$

12.2.2 Fourier Equation

The extension of Newton's law of cooling leads to Fourier's law of heat conduction. The heat flux is given by

$$\mathbf{J} = -k\nabla T \tag{12.12}$$

and the thermodynamic force is

$$\nabla T^{-1}. \tag{12.13}$$

Again we find the flux not to be proportional to the force. We insert the flux into the balance expression

$$\rho c(\partial T/\partial t) + \nabla \cdot \mathbf{J} \tag{12.14}$$

set to zero, and obtain Fourier' law

$$\partial T/\partial t = \lambda \nabla^2 T, \tag{12.15}$$

where $\lambda = k/\rho c$ is the thermal diffusivity. The entropy production of the whole system is a functional of the temperature field $T(\boldsymbol{r})$ and is

$$\sigma[T(\mathbf{r})] = \int \mathbf{J} \cdot \nabla T^{-1} \mathrm{d}\mathbf{r} = k \int \frac{[\nabla T(\mathbf{r})]^2}{[T(\mathbf{r})]^2} \mathrm{d}\mathbf{r} \geq 0. \tag{12.16}$$

To see if the entropy production rate has an extremum for the stationary state corresponding to a constant heat flux $\mathbf{J}_0$ we evaluate the functional derivative with respect to the temperature field

$$\frac{\delta}{\delta T(\mathbf{r}')} \sigma[T(\mathbf{r})] = 2k \int \left[\left(\frac{1}{T(\mathbf{r})} (\nabla_{\mathbf{r}} \delta(\mathbf{r} - \mathbf{r}')) \cdot (\nabla \ln\, T(\mathbf{r})) - \right.\right.$$

$$\left.\left. \frac{\delta(\mathbf{r} - \mathbf{r}')}{T(\mathbf{r})} (\nabla \ \ln \ T(\mathbf{r}))^2 \right) \right] \mathrm{d}\mathbf{r} = 2k \left[\frac{(\nabla T(\mathbf{r}'))^2}{(T(\mathbf{r}'))^3} - \frac{\nabla^2 T(\mathbf{r}')}{(T(\mathbf{r}'))^2} \right] \tag{12.17}$$

For a stationary state Fourier's equation (12.15) reduces to Laplace's equation

$$\nabla^2 T_{\mathrm{st}}(\mathbf{r}) = 0. \tag{12.18}$$

We insert the stationarity condition $\mathbf{J}_0 = -k\nabla T_{\mathrm{st}}\,(r)$ into (16, 17) to obtain

$$\sigma[T_{\mathrm{st}}(\mathbf{r})]|_{\mathrm{st}} = \frac{|\mathbf{J}_0|^2}{k} \int \frac{\mathrm{d}\mathbf{r}}{|T_{\mathrm{st}}(\mathbf{r})]^2} = \begin{cases} = 0 \text{ if } |\mathbf{J}_0| = 0 \\ > 0 \text{ if } |\mathbf{J}_0| > 0 \end{cases} \tag{12.19}$$

and

$$\frac{\delta}{\delta T(\mathbf{r}')} \sigma[T(\mathbf{r})]|_{\mathrm{st}} = \frac{2|\mathbf{J}_0|^2}{k(T_{\mathrm{st}}(\mathbf{r}'))^3} = \begin{cases} = 0 \text{ if } |\mathbf{J}_0| = 0 \\ > 0 \text{ if } |\mathbf{J}_0| > 0. \end{cases} \tag{12.20}$$

These equations show that the entropy production rate is an extremum if and only if the stationary state is the state of equilibrium.

12.3 Exact Solution for Chemical Reactions

Consider a general reaction network with ideal mass-action laws of kinetics

$$\sum_{u=1}^{S_a} \alpha_{uw}^{+} A_u + \sum_{u=1}^{S_x} \beta_{uw}^{+} X_u \rightleftharpoons \sum_{u=1}^{S_a} \alpha_{uw}^{-} A_u + \sum_{u=1}^{S_x} \beta_{uw}^{-} X_u, \tag{12.21}$$

where the forward and backward extensive reaction rates are

$$r_w^{\pm}(\mathbf{a}, \mathbf{x}) = V k_w^{\pm} \left[\prod_{u=1}^{S_a} (a_u)^{\alpha_{uw}{}^{\pm}} \right] \left[\prod_{u=1}^{S_x} (x_u)^{\beta_{uw}{}^{\pm}} \right]. \tag{12.22}$$

In these equations, $A_u, u = 1, \ldots, S_a$ are stable species with concentrations $a_u, u = 1, \ldots, S_a$ kept constant by interactions with a set of reservoirs connected to the system, and $X_u, u = 1, \ldots, S_u$ are reaction intermediates with variable concentrations $x_u, u = 1, \ldots, S_x$. V is the volume of the system. The system can be kept away from equilibrium by controlling the concentrations of the stable species.

The entropy production rate is the sum of the product of the net flux times the affinity for each elementary reaction step (compare with the form of (1.21))

$$\sigma(\mathbf{a}, \mathbf{x}) = k_{\mathrm{B}} \sum_{w=1}^{R} [r_w^{+}(\mathbf{a}, \mathbf{x}) - r_w^{-}(\mathbf{a}, \mathbf{x})] \ln \left[\frac{r_w^{+}(\mathbf{a}, \mathbf{x})}{r_w^{-}(\mathbf{a}, \mathbf{x})} \right] \geq 0. \tag{12.23}$$

The entropy production depends only on concentrations but not explicitly on time. The differential of the entropy production can be evaluated with the use of (12.23)

$$\delta\sigma(\mathbf{a}, \mathbf{x}) = \sum_{u=1}^{S_x} \delta \ln x_u \frac{\partial}{\partial \ln x_u} \sigma(\mathbf{a}, \mathbf{x}) = \sum_{u=1}^{S_x} \tilde{r}_u(\mathbf{a}, \mathbf{x}) \delta \ln x_u + \sum_{u=1}^{S_x} \mathscr{B}_u(\mathbf{a}, \mathbf{x}) \delta \ln x_u, \tag{12.24}$$

where the net reaction rates of the species are

$$\tilde{r}_u(\mathbf{a}, \mathbf{x}) = \sum_{w=1}^{R} [r_w^{+}(\mathbf{a}, \mathbf{x}) - r_w^{-}(\mathbf{a}, \mathbf{x})](\beta_{uw}^{+} - \beta_{uw}^{-}) \tag{12.25}$$

and

$$\mathcal{B}_u(\mathbf{a}, \mathbf{x}) = \sum_{w=1}^{R} [\beta_{uw}^{+} r_w^{+}(\mathbf{a}, \mathbf{x}) - \beta_{uw}^{-} r_w^{-}(\mathbf{a}, \mathbf{x})] \ln \left[\frac{r_w^{+}(\mathbf{a}, \mathbf{x})}{r_w^{-}(\mathbf{a}, \mathbf{x})} \right]. \tag{12.26}$$

For a stationary state $\mathbf{x} = \mathbf{x}^{\mathrm{st}}$ the net reaction rates of the active species equals zero and the first sum in the second of the equations in (12.24) is zero. The second sum in the second of equations in (12.24) is however in general not zero. It is zero only at thermodynamic equilibrium where detailed balance holds

$$r_w^+ (\mathbf{a}, \mathbf{x}) = r_w^- (\mathbf{a}, \mathbf{x}). \tag{12.27}$$

We now select a very simple example of a reaction mechanism consisting of two elementary reactions

$$\nu A \rightleftharpoons X\nu, \ \nu X \rightleftharpoons \nu B \tag{12.28}$$

to show that the entropy production never has an extremum except at equilibrium. A few counter examples suffice to negate the principle of minimum entropy production for chemical reactions [2,3]. For the reaction (12.28) there is a single stable stationary state for which we have

$$x = x^{\mathrm{st}} = \left\{ \frac{1}{2}[(a)^\nu + (b)^\nu] \right\}^{1/\nu}. \tag{12.29}$$

For $a \neq b$ the stationary state is a non-equlibrium state, and for $a = b = x^{\mathrm{st}}$ the stationary state is one of thermodynamic equilibrium. The variation of the entropy production rate at $x = x^{\mathrm{st}}$ is

$$\delta\sigma(a, b, x^{\mathrm{st}}) = (\delta \ \ln \ x^{\mathrm{st}}) V k_{\mathrm{B}} \nu k \frac{1}{2}[(a)^\nu + (b)^\nu] \ln \left(\frac{[(a)^\nu + (b)^\nu]^2}{4a^\nu b^\nu} \right) \tag{12.30}$$

From the algebraic inequality

$$\frac{[(a)^\nu + (b)^\nu]^2}{4a^\nu b^\nu} = \begin{cases} 1 \text{ for } a = b \\ > 1 \text{ for } a \neq b \end{cases} \tag{12.31}$$

it follows that

$$\frac{\delta\sigma(a, b, x^{\mathrm{st}})}{\delta \ \ln \ x^{\mathrm{st}}} = \begin{cases} 0 \text{ for } a = b \\ > 0 \text{ for } a \neq b \end{cases} \tag{12.32}$$

and the entropy production rate has an extremum if and only if the system is at thermodynamic equilibrium. For non-equilibrium stationary states, no matter how close to equilibrium, the entropy production rate does not have an extremum.

We have shown explicitly and without approximation that for two cases of irreversible processes the rate of entropy production has no extremum at stationary states neither near nor far from euilibrium. If the flux of any transport process is strictly proportional to the force then the entropy production is the square of either and trivially has an extremum, a minimum, at a stationary state.

12.4 Invalidity of the Principle of Minimum Entropy Production

In 1987 we were concerned with the validity of the so-called principle of minimum entropy production rate [4, 5]. In the first article we showed by expansion of the entropy production the general invalidity of the principle. Once the entropy production rate is expanded in the affinity, the deviation from equilibrium, then two operations are required (1). the differentiation of the entropy production rate with respect to temperature and (2). the termination of the series expansion in the affinity to simulate the requirement 'close to equilibrium'. The problem arises with the fact that these two operations do not commute. Only if operation 2 preceeds 1, an incorrect procedure, then the dissipation shows an extremum at a stationary state. Only the incorrect procedure leads to a 'principle'. If operation 1 preceeds 2 then the dissipation has no extremum at a stationary state, the same result as obtained in Sects. 12.2 and 12.3 without any approximations.

Glansdorff and Prigogine [3] stated the principle in the following way: '... if the steady states occur *sufficiently close to equilibrium states* they may be characterized by an extremum principle according to which *the entropy production has its minimum value at the steady state compatible with the prescribed conditions* (constraints).'

Over the years much has been made of the principle of minimum entropy production. A few quotes will suffice.

Lehninger in the 1975 edition of his well-known text on biochemistry [6] states '.. at least two general attributes of open systems have considerable significance in biology ... the most important implication is this: in the formalism of nonequilibrium thermodynamics, the steady state, which is characteristic of all smoothly running machinery, may be considered to be the orderly state of an open system, the state in which the rate of entropy production is at a minimum and in which the system is operating with maximum efficiency under the prevailing conditions'.
Clearly wrong. A broader statement was made by Katchalsky [7]: 'This remarkable conclusion ... sheds new light on the 'wisdom of living organisms'. Life is a constant struggle against the tendency to produce entropy by irreversible processes. The synthesis of large, information rich macromolecules, the formation of intricately structured cells, the development of organization, all these are powerful anti-entropic forces. But since there is no possibility of escaping the entropic doom imposed on all natural phenomena under the Second Law of Thermodynamics, living systems choose the least evil–they produce entropy at a minimum rate by maintaining a steady state.'
Colorful but wrong. Voet and Voet, in a widely used text on 'Biochemistry' [8] state: "Ilya Prigogine, a pioneer in the development of irreversible thermodynamics, has shown that a steady state produces the maximum amount of useful work for a given energy expenditure under the prevailing conditions. *The steady state of an open system is therefore a state of maximum efficiency.*"

Wrong again on several counts, see Chap. 13 and later.

12.5 Invalidity of the 'Principle of Maximum Entropy Production'

This principle states that irreversible processes proceed in a direction which produces maximum entropy production. If there is a choice then the path with the highest entropy production has the fastest rate. This principle is much less well known, yet there are hundreds of articles on this subject and several reviews [9]. For chemistry, and hence for biology, the principle is generally invalid. The rates of reactions are governed by the Gibbs free energy of activation, not by the rate of entropy production. There is in general no relation between thermodynamic quantities, such as the rate of entropy production, the Gibbs free energy, etc, and quantities determining the rates of reactions. There may some empirical relations but they are of limited use. There is experimental evidence against this principle [10].

See Chap. 2, (2.40) for an expression of equistablity in terms of an integral over time of the species-specific dissipation, (2.26). It may appear to be some connection with the 'principle' under discussion, but there is none.

In Sect. 8.4.1, we discuss a Rayleigh–Benard experiment, see [11] in Chap. 8, which is a clear refutation of this 'principle' in the field of transport processes and hydrodynamics.

12.6 Editorial

If a theory is proven wrong, or a better theory has been offered, how long does the process of adaptation by the scientific community take? The longer the old theory has been believed to be correct, the longer the time of adaptation of a new and better theory. Scientists seldom change their mind, they cease to practice science and they die off. (A rephrasing of a thought expressed by Max Planck.) Younger scientists not steeped in the old theory tend to adapt. A point in case: The second edition of 'Physical Chemistry', a widely used text by Noyes and Sherrill published in 1938, makes no mention of quantum mechanics, nearly 50 years after Planck and 12 years after Schrodinger and Heisenberg. Several books on quantum mechanics, including Pauling and Wilson's 'Introduction to Quantum Mechanics' had appeared by 1935.

Acknowledgement. This chapter is based on part of [1].

References

1. J. Ross, M.O. Vlad, J. Phys. Chem. A. 109, 10607–10612 (2005)
2. I. Prigogine, Bull. Cl. Sci., Acad. R. Belg. **31**, 600–606 (1945)
3. P. Glansdorff, I. Prigogine, *Thermodynamic Theory of Structure, Stability, and Fluctuations* (Wiley, New York, 1971)

4. K.L.C. Hunt, P.M. Hunt, J. Ross, Phys. A., **147**, 48–60 (1987)
5. K.L.C. Hunt, P.M. Hunt, J. Ross, Phys. A. **154**, 207–211 (1988)
6. A.L. Lehninger, *Biochemistry* (Worth, New York, 1975), pp. 412–413. The statement is omitted from a subsequent text. A.L. Lehninger, *Principles of Biochemistry* (Worth, New York, 1982)
7. A. Katchalsky, in *Modern Science and Technology*, ed. by R. Colborn, (D. Van Nostrand, New York, 1965)
8. D. Voet, J.G. Voet, *Biochemistry*, 3rd ed.; Wiley, New York, 2004, 574–579
9. L.M. Martyushev, V.D. Seleznev, Phys. Reports **426**, 1–45 (2006)
10. B. Andresen, E.C. Zimmerman, J. Ross, J. Chem. Phys., **81**, 4676–4677 (1984)

13

Efficiency of Irreversible Processes

13.1 Introduction

A chemical reaction with a Gibbs free energy change less than zero (negative) can proceed spontaneously, irreversibly, and can produce work. If the reaction is run reversibly, then the maximum work, other than pV work, is ΔG, for the reaction as written. If the reaction is run irreversibly then there is some entropy production and some work may be done, but less than ΔG. If no work is done then the rate of entropy production is

$$T\frac{\mathrm{d}S}{\mathrm{d}t} = \Delta G(\text{rate}), \tag{13.1}$$

where 'rate' denotes the rate of the reaction. It is instructive to derive this equation: The definition of the Gibbs free energy is

$$G = H - TS, \tag{13.2}$$

where all quantities refer to the system. Hence we have

$$\frac{\mathrm{d}G}{\mathrm{d}t} = \frac{\mathrm{d}H}{\mathrm{d}t} - T\frac{\mathrm{d}S}{\mathrm{d}t} \tag{13.3}$$

at constant T. At constant pressure the enthalpy term becomes

$$\frac{\mathrm{d}H}{\mathrm{d}t} = \frac{\mathrm{d}Q_p}{\mathrm{d}t} = -T\frac{\mathrm{d}S_{\text{surr}}}{\mathrm{d}t} \tag{13.4}$$

so that

$$\frac{\mathrm{d}G}{\mathrm{d}t} = -T\frac{\mathrm{d}S_{\text{surr}}}{\mathrm{d}t} - T\frac{\mathrm{d}S_{\text{system}}}{\mathrm{d}t} \tag{13.5}$$

or

$$\frac{\mathrm{d}G_{\text{system}}}{\mathrm{d}t} = -T\frac{\mathrm{d}S_{\text{universe}}}{\mathrm{d}t}. \tag{13.6}$$

By convention the sum of an entropy change in the system plus that of the surroundings is called the entropy change of the universe, and the lhs of (13.6) is another way of writing the rhs of (13.1).

As an example consider a reaction run in an electro-chemical cell: There are two half cells with an electrode in each of the two half cells. If the electrodes are shorted (connected with a wire) then the reaction produces no work, only dissipation (heat). If a potential is applied externally across the electrodes, equal and opposite to that generated by the electro-chemical cell, then the reaction is at equilibrium. A differential change in the applied potential produces an infinitesimal electron current and an infinitesimal, reversible amount of electrical work. For a finite amount of electrical work produced reversibly we have

$$\Delta G = -ENF = W_X, \tag{13.7}$$

where E is the cell potential, N the number of electrons transferred from one electrode to the other for the cell reaction as written and F is the Faraday constant, the number of Coulombs per mole of electrons. If the applied potential is less than the cell potential, then less than the maximum work is done by the cell. The difference between the cell potential and the applied potential, times the amount of charge transferred by unit time, equals the product of the temperature times the rate of entropy production, that is the dissipation.

The power output of an electro-chemical cell (or a battery) is the product of the applied voltage times the current. A reversible electro-chemical cell has zero current and hence zero power output, and zero entropy production. Power output requires an irreversible process run in finite time; therefore power output is concurrent with entropy production (dissipation). One use of the term 'efficiency' of a process, such as a reaction, is in reference to energy transduction, for example what part of ΔG of a reaction is used to produce work; there are other uses of this term.

13.2 Power and Efficiency of Heat Engines

We begin with an analysis of heat engines, which are devices, run in a cycle, that accept heat at a high temperature, do some work, and reject heat at a lower temperature. The standard example is the reversible Carnot engine which runs in a cycle starting, say, at V_1 (Fig. 13.1) and the upper reservoir temperature T_1, proceeding isothermally at T_1 to V_2, then adiabatically to V_3 and T_2, then isothermally at the lower reservoir temperature T_2 to V_4, and finally returning adiabatically to the starting point at T_1 and V_1.

On the first isothermal branch a quantity of heat, Q_1, flows reversibly from the heat reservoir to the engine, both being a T_1; similarly, on the second isothermal branch a quantity of heat Q_2 flows from the engine to the heat reservoir, both at T_2. A total amount of work W is done in one cycle. For the isothermal steps we have the entropy changes

$$\Delta S_1 = \frac{Q_1}{T_1} \qquad \text{and} \qquad \Delta S_2 = \frac{Q_2}{T_2} \tag{13.8}$$

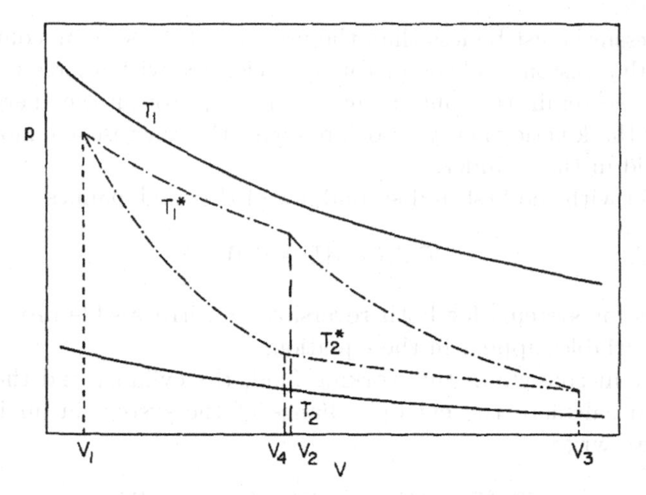

Fig. 13.1. Pressure–volume diagram of an isothermal heat cycle, from [1]

or, for the cycle

$$\frac{Q_1}{T_1} + \frac{Q_2}{T_2} = 0. \tag{13.9}$$

Since for the cycle we have

$$\Delta E = 0 = W + Q_1 + Q_2 \tag{13.10}$$

then

$$-\frac{W}{Q_1} = 1 - \frac{T_2}{T_1} \tag{13.11}$$

the well-known expression for the efficiency of a reversible Carnot engine, that is the maximum work available per cycle for a given heat input Q_1 and the given reservoir temperatures.

The reversible Carnot engine has no power output, since the reversible work done takes an infinite time. The power is the work done in a finite time, and hence here is zero. To achieve power output [1] there must be some spontaneous, natural, irreversible process [2, 3]. We shall assume that heat flows spontaneously from the reservoir at T_1 to the heat engine at $T_1{}^*$, see Fig. 13.1, according to a simple linear rate law

$$\mathrm{d}Q_1/\mathrm{d}t = \alpha(T_1 - T_1^*) \tag{13.12}$$

during the isothermal expansion from V_1 to V_2. The cycle continues with an adiabatic expansion from V_2 to V_3 and $T_2{}^*$; next comes an isothermal compression to V_4, where the heat engine at $T_2{}^*$ is connected to the heat reservoir at T_2 and heat flows at a non-zero rate according to an equation similar to (13.12). The final step in the cycle is an adiabatic compression back to V_1 and $T_1{}^*$. During the isothermal expansion, the first step in the cycle, the external pressure must be less than the pressure of the system, contained in a cylinder with a piston, and the piston is accelerated with a force proportional to $p - p_{\text{ext}}$; we retain the kinetic energy of the piston in the energy balance but neglect the kinetic energy associated with the macroscopic motion of the working fluid in the cylinder.

We begin with the first and second law of thermodynamics

$$T\,\mathrm{d}S = \mathrm{d}E + p\,\mathrm{d}V, \tag{13.13}$$

which holds for systems for both reversible and irreversible processes, since only state variables appear in the equation.

Our system consists of the working fluid, the cylinder and the piston of mass m. To calculate the entropy change of the system in an irreversible expansion we write

$$\begin{aligned} T\,\mathrm{d}S &= \mathrm{d}E + p_{\mathbf{ex}}\mathrm{d}V + (p - p_{\mathbf{ex}})\mathrm{d}V \\ &= \mathrm{d}Q - \mathrm{d}K + (p - p_{\mathbf{ex}})\mathrm{d}V, \end{aligned} \tag{13.14}$$

where $\mathrm{d}K$ is the kinetic energy change of the piston. The piston satisfies Newton's second law

$$p - p_{\mathrm{ex}} = m\frac{\mathrm{d}^2 V}{\mathrm{d}t^2} \tag{13.15}$$

for a piston of unit area. For the kinetic energy we have

$$K = \frac{m}{2}\left(\frac{\mathrm{d}V}{\mathrm{d}t}\right)^2,$$

$$\mathrm{d}K = m\frac{\mathrm{d}^2 V}{\mathrm{d}t^2}\mathrm{d}V \tag{13.16}$$

and hence we find

$$\mathrm{d}S = \mathrm{d}Q/T. \tag{13.17}$$

By including the piston in the system we see that the entropy change of the system can be written in the same form for reversible and irreversible processes. With inclusion of the piston in the system we insure that on expansion, reversible or irreversible, there is no dissipation and hence (13.17) applies in both cases.

We turn next to the calculation of the rates of all the steps in the irreversible cycle. For the first isothermal step heat flows from the reservoir to the system and may change the internal energy of the system and the kinetic energy of the piston, and may be used to produce work in the surroundings. We may write

$$\frac{\mathrm{d}Q_1}{\mathrm{d}t} = \frac{\mathrm{d}E}{\mathrm{d}t} + \frac{\mathrm{d}K}{\mathrm{d}t} + p_{\mathrm{ex}}\frac{\mathrm{d}V}{\mathrm{d}t}, \tag{13.18}$$

If the working fluid is taken to be an ideal gas then $\mathrm{d}E/\mathrm{d}t = 0$ for the isothermal expansion so that

$$\frac{\mathrm{d}Q_1}{\mathrm{d}t} = m\frac{\mathrm{d}^2V}{\mathrm{d}t^2}\frac{\mathrm{d}V}{\mathrm{d}t} + p_{\mathrm{ex}}\frac{\mathrm{d}V}{\mathrm{d}t}. \tag{13.19}$$

On combining (13.19), (13.20) and (13.12) we obtain

$$\begin{aligned}\frac{\mathrm{d}Q_1}{\mathrm{d}t} &= \alpha(T_1 - T_1^*) = p\frac{\mathrm{d}V}{\mathrm{d}t} \\ &= \frac{RT_1^*}{V}\frac{\mathrm{d}V}{\mathrm{d}t},\end{aligned} \tag{13.20}$$

where in the second line we have substituted the equation of state for one mole of an ideal gas. The equation of motion for the volume of the system is therefore

$$\mathrm{d}V/\mathrm{d}t = V/f_1, \tag{13.21}$$

with

$$f_1 \equiv \frac{R}{\alpha}\frac{T_1^*}{T_1 - T_1^*}. \tag{13.22}$$

The time necessary to complete the isothermal expansion from V_1 to V_2 is

$$t_1 = f_1 \ln(V_2/V_1) \tag{13.23}$$

and the heat transferred is

$$\begin{aligned}Q_1 &= \int_0^{t_1} \alpha(T_1 - T_1^*)\mathrm{d}t = \alpha t_1(T_1 - T_1^*) \\ &= RT_1^* \ln(V_2/V_1).\end{aligned} \tag{13.24}$$

For the first adiabatic branch the system is decoupled from the heat reservoir at T_1; $\mathrm{d}Q$ is zero and from (13.17) we see that the system undergoes a reversible adiabatic expansion from V_2 to V_3 at temperature T^*, which, for an ideal gas, gives

$$C_v\frac{\mathrm{d}T^*}{\mathrm{d}t} + \frac{RT^*}{V}\frac{\mathrm{d}V}{\mathrm{d}t} = 0. \tag{13.25}$$

The external pressure in the adiabatic processes is undetermined; we choose p_{ext} to have the form

$$p_{\mathrm{ex}} = \frac{RT^*}{V} - m\frac{V}{f_1^2}, \tag{13.26}$$

so that the equation of motion becomes identical to (13.17); the upper bounds on the power and efficiency of the heat engine are independent of the choice of this process. With (13.17) we find that the time necessary for this adiabatic expansion is

$$t_2 = f_1 \ln\frac{V_3}{V_2} = \frac{f_1}{(\gamma - 1)}\ln\frac{T_1^*}{T_2^*}, \tag{13.27}$$

where γ is the ratio of the heat capacities C_p/C_v and we have used the relation

$$T_1^* V_2^{\gamma-1} = T_2^* V_3^{\gamma-1} \tag{13.28}$$

for an adiabatic process of an ideal gas. The kinetic energy of the piston is convertible into work.

For the next isothermal and adiabatic compression steps we proceed in a, respectively, parallel way. The equation for the isothermal compression is

$$\mathrm{d}V/\mathrm{d}t = V/f_2 \tag{13.29}$$

with

$$f_2 \equiv \frac{R}{\alpha} \frac{T_2^*}{T_2^* - T_2}; \tag{13.30}$$

the time for the isothermal compression is

$$t_3 = f_2 \ln(V_3/V_4) \tag{13.31}$$

and the heat transferred during the second isothermal branch is

$$Q_2 = -RT_2^* \ln(V_3/V_4). \tag{13.32}$$

The time for the adiabatic compression (compare with (13.27)) is

$$t_4 = -f_2 \ln \frac{V_1}{V_4} = \frac{f_2}{(\gamma - 1)} \ln \frac{T_1^*}{T_2^*}. \tag{13.33}$$

The total work done by the system, including the conversion of the kinetic energy of the piston, is for one complete cycle obtained from the first law: since $\Delta E = 0$ for the cycle we have

$$W = Q_1 + Q_2 \tag{13.34}$$

or from (13.24) and (13.32)

$$W = RT_1^* \ln \frac{V_2}{V_1} - RT_2^* \ln \frac{V_3}{V_4}. \tag{13.35}$$

The second and fourth steps are effectively reversible adiabatics and hence we have

$$\begin{aligned} T_1^* V_1^{\gamma-1} &= T_2^* \; V_4^{\gamma-1}, \\ T_1^* V_2^{\gamma-1} &= T_2^* \; V_3^{\gamma-1}, \end{aligned} \tag{13.36}$$

where

$$\gamma - 1 + (R/C_v) = C_p/C_v; \tag{13.37}$$

thus

$$V_2/V_1 = V_3/V_4 \tag{13.38}$$

and

$$V_2 = V_3 \left(\frac{T_2^*}{T_1^*}\right)^v \tag{13.39}$$

where $\nu = 1/(\gamma - 1)$. With these identities we can change (13.35) to

$$W = R(T_1^* - T_2^*)\left(\ln\frac{V_2}{V_1} + \nu \ln\frac{T_2^*}{T_1^*}\right). \tag{13.40}$$

We calculate the average power per cycle by first finding the total cycle time, t_{tot}, from (13.23), (13.27), (13.31), (13.33)

$$t_{\mathrm{tot}} = f_1 \ln\frac{V_2}{V_1} + f_2 \ln\frac{V_2}{V_1} + \nu(f_1 + f_2)\ln\frac{T_1^*}{T_2^*}. \tag{13.41}$$

Hence the power output is

$$P = \frac{W}{t_{\mathrm{tot}}} = \frac{\alpha}{\ln V_3/V_1}\frac{(T_1^* - T_2^*)\left(\ln\frac{V_3}{V_1} + \nu\ln\frac{T_2^*}{T_1^*}\right)}{\frac{T_1^*}{T_1 - T_1^*} + \frac{T_2^*}{T_2^* - T_2}}, \tag{13.42}$$

which is an explicit function of the compression ratio, V_3/V_1, the ratio of the maximal to minimal volume of the working fluid in the heat engine.

In Fig. 13.2, taken from [1], we plot the average power output divided by the maximum power output as a function of the difference between the temperatures of the engine during the isothermal steps of the cycle.

Curzon and Ahlborn [2], in their original work on this subject, argue that the power output on this plot must have a maximum: for an explanation see the caption to Fig. 13.2. The efficiency of the engine is defined as W/Q_1 and thus we have

$$\eta = \frac{W}{Q_1} - 1 + \frac{Q_2}{Q_1} - 1 - \frac{T_2^*}{T_1^*}, \tag{13.43}$$

where we have used (13.24) and (13.32). To obtain the maximum power we need to differentiate the expression of the power output (13.42), which is given in detail in Appendix A of [1]. In the limit of large compression ratio we obtain for the efficiency at maximum power

$$\eta_m \leq 1 - \sqrt{T_2/T_1}. \tag{13.44}$$

Consider an example for a heat engine where the high temperature reservoir is at T_1, say 1,000 °K and the low temperature reservoir at T_2, say 500 °K. The efficiency of the reversible Carnot engine, (13.11), is 0.5, whereas the efficiency of an heat engine with power output, at maximum power, (13.44), is about 0.3. Due to the requirement of power output there is a 40% drop in the efficiency of the heat engine! This is the main lesson of this chapter. The analysis is idealized but the makes the point, that achieving power output necessitates losses in efficieny.

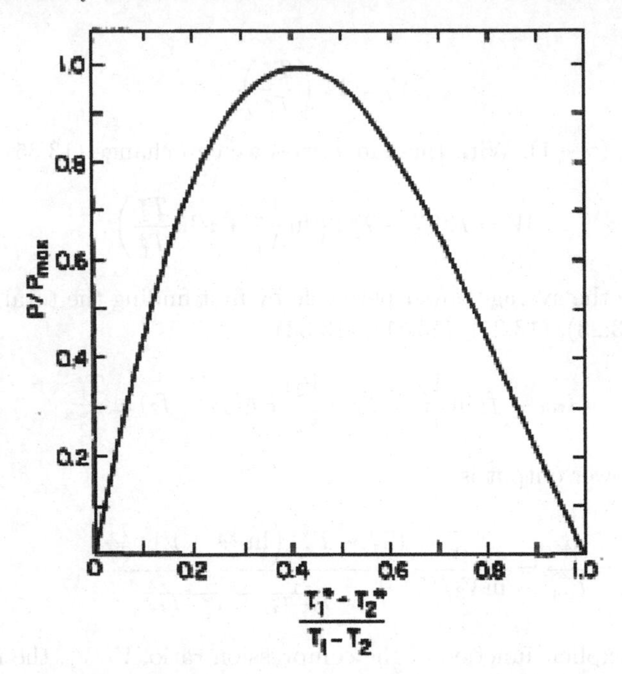

Fig. 13.2. The average power output (normalized by $P_{\max}$) as a function of the difference between the engine's temperatures during the isothermal process. The following parameters were used in this figure: $T_2/T_1 = 0.3$, $\ln(V_3/V_1) = 3$, and $(T_1^*/T_1) = T_2^*/(2T_2^* - T_2)$, which corresponds to maximization with respect to T_1^*. The power vanishes when $T_1^* = T_1$ or $T_2^* = T_2$ because the cycle time is then infinite. It vanishes again when $T_1^* = T_2^*$ because then no work is done by the system. Thus, the power output must have a maximum at some intermediate temperatures

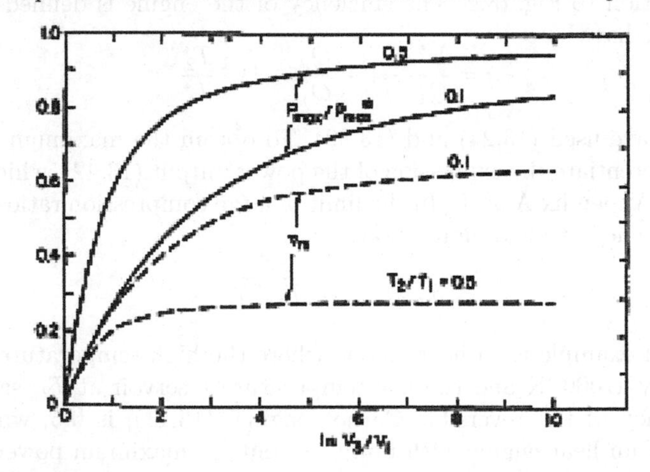

Fig. 13.3. The graph is explained in the text, from [1]

In Fig. 13.3 we show values of the maximum power output, normalized by the maximum power output at infinite compression ratio, and the efficiency at the maximum power output for two values of the ratio $(T_1/T_2) = 0.5$ and 0.1 as a function of the compression ratio. The upper bounds are obtained in the infinite compression ratio, and are independent of the choice of p_{ex} in the adiabatic steps.

In this limit the time spent on the adiabatics is negligible compared to the total time for one cycle and thus (13.44) is valid for any cycle with two isothermal processes. Moreover, in the limit of high compression ratio the properties of any working fluid approach those of an ideal gas during most of the cycle and hence (13.44) is valid regardless of the working fluid of the heat engine.

In [1], other proofs are given including one to show that the maximum power cycle is one with two isothermal steps.

In [4] there is an analysis of time-dependent thermodynamic systems by obtaining the entropy production in terms of the relevant relaxation times in the system. The various physically possible limits of these relaxation times and their ratios leads to a classification into reversible, both quasistatic and otherwise, and irreversible processes. In one of these limits it is possible for a reversible process not to be quasistatic, but this limit is not physically interesting since for thermal conduction it would require infinite thermal conductivity. Hence our statement that power output requires irreversible processes is here substantiated.

In the next Chap. 14, there is a presentation of 'Finite Time Thermodynamics', by R.S. Berry, which is another way of formulating thermodynamics for systems not at equilibrium.

Acknowledgement. Section 13.2 of this chapter has been taken, with some editing and rewording from [1].

References

1. D. Gutkowicz-Krusin, I. Procaccia, J. Ross, J. Chem. Phys. **69**, 3898–3906 (1978)
2. F.L. Curzon, B. Ahlborn, J. Phys. **43**, 22–24 (1975). Curzon and Ahlborn were the first to publish on this subject. Their derivation is limited to infinite compression ratio, whereas we present an analysis valid for any compression ratio
3. (a) B. Andresen, R.S. Berry, A. Nitzan, P. Salamon, Phys. Rev. A **15**, 2086–2093 (1977) (b) B. Andresen, R.S. Berry, P. Salamon, Phys. Rev. A **15**, 2094–2102 (1977) (c) B. Andresen, R.S. Berry, P. Salamon, J. Chem. Phys. **66**, 1571–1577 (1977)
4. V. Fairen, M.D. Hatlee, J. Ross, J. Phys. Chem. **80**, 70–73 (1982)

14

Finite-Time Thermodynamics

R. Stephen Berry

University of Chicago

14.1 Introduction and Background

Traditional thermodynamics evolved from Carnot's introduction of the concept of the ideal reversible process, a process that would proceed infinitely slowly. Precisely because of that constraint, such a process would incur none of the losses of friction or other kinds of dissipation that result from real-time operation. The concepts of thermodynamic potentials such as the free energies provide limits of performance based on exactly those reversible processes as the standards of comparison with real processes. Onsager and then others showed that one can say useful things about irreversible processes, especially those near equilibrium. An entire field of engineering thermodynamics grew out of the concept of 'local thermal equilibrium' or LTE, in which one can describe a large system such as a flow process in terms of how the system changes as it moves, in effect, through a succession of steps which may be near or quite far from true thermodynamic equilibrium but that can be assigned local temperatures.

Here we introduce another approach to dealing with non-equilibrium processes. The goal here is ultimately finding ways to improve the efficiency of energy use for various technologies, by using the analogues of traditional thermodynamic potentials for processes whose very definitions constrain them to operate at non-zero rates or in finite time. These analogues determine the ideal limits of performance for the time- or rate-constrained systems, and offer an alternative to the methods that are given in the other chapters of this book, especially those of Chap. 8. We first examine the background that motivated the introduction of this approach, then review the fundamental theorems that justify their existence, move on to show how one can find those finite-time limits of performance, and finally, show some examples of how one can optimize particular processes to approach as closely as possible to those limits. Since we are going beyond the equilibrium domain, we must include in the definition of any finite-time process the inherent, unavoidable irreversibilities, as was discussed in Chap. 12. In this approach, we deal only with the traditional mean values of the thermodynamic quantities and not with fluctuations.

The motivation for what became known as 'Finite-Time Thermodynamics' was a search for ways to identify especially attractive targets for improving efficiency of energy use. That, in turn, had been motivated by the author's concern about the severe air pollution that afflicted many American and other cities for many years, through the 1960s. It seemed that one should try to go beyond addressing symptoms by, for example, putting precipitators onto sources of emission, and try to reduce the levels of pollutants by reducing the amounts of energy whose consumption was generating those pollutants. Empirical comparisons of actual energy use, step-by-step, in many kinds of processes, with the ideal thermodynamic limits for those same processes proved to be a very useful guide. Processes whose energy and free energy consumptions differed significantly from the thermodynamic limits were good targets for technological improvements. However one observer challenged this work: 'Why are you comparing the actual use with the ideal, reversible limit? Who would order a car from a manufacturer who made his products infinitely slowly?'

This question led to a very productive *scientific* question: 'What would be the necessary and sufficient conditions for the existence of analogues of the thermodynamic potentials, for processes constrained to operate in finite time?' Answering this question was almost the first step in the development of the subject. However in fact two prior steps really began the subject. The first was a study by Curzon and Ahlborn [1] introduced in Chap. 13, which examined the conditions under which the Carnot engine would operate to maximize *power*, rather than maximizing *efficiency*, which had been the traditional quantity to optimize ever since Carnot introduced the engine operating on a cycle of two isothermal steps separated by two adiabatic steps, all precisely reversible. The Curzon–Ahlborn engine, like any other, of course must operate at a non-zero rate or it produces no power at all. The result of that study was a striking conclusion that the efficiency $\eta_{\max P}$ of the Carnot-type engine operating to produce maximum power $P_{\max}$ occurs when there is a specific relation between the two temperatures $T_{\rm H}$ and $T_{\rm L}$ between which the process operates: $\eta_{\max P} = 1 - [T_{\rm L}/T_{\rm H}]^{1/2}$. This is, of course, very reminiscent of the efficiency $\eta_{\rm rev}$ of the reversible Carnot engine, which is $\eta_{\rm rev} = 1 - [T_{\rm L}/T_{\rm H}]$. Curzon and Ahlborn showed that some real power plants in fact operate with efficiencies rather close to those corresponding to maximum power production. Chapter 13 introduced the Carnot engine and its variation to allow irreversible expansion and heat transfer, in order to do work and incorporate the constraint of finite-time operation.

The second step historically in the approach discussed here was a sort of test case, the analysis of a simple model system that consisted of a Carnot cycle that operated in short, finite-time steps [2]. In this work, the system operates through a series of small, discrete steps in which the pressure changes discontinuously and the system is connected to its heat reservoirs by finite heat conductances. The results gave the values and conditions for maximum effectiveness, the ratio of the work actually done, per cycle, to the total change of availability, the maximum work that could be extracted from the cycle. They also gave the maximum efficiency, the maximum power per cycle, the optimal period per cycle and the rate at which that maximum power is delivered.

Following that analysis of a model problem came the rigorous basis of what became 'Finite-Time Thermodynamics,' the theorems giving necessary and sufficient conditions for the existence of potential-like quantities for finite-time processes and a first algorithm to evaluate such potential-like functions [3]. This approach relies on relating thermodynamics to the formalism of classical mechanics, in the spirit of Chap. 3 but uses somewhat different mathematical tools. The essence is the construction of an exact differential for the process of interest, a quantity that satisfies the constraints of that process but is independent of the path, so long as it satisfies those constraints. The two quantities whose limits one wants such potentials to provide are of course heat exchanged and work done. In traditional thermodynamics, the enthalpy H is a potential for heat exchanged in a process at constant pressure; the Gibbs free energy is its counterpart for work exchanged at constant temperature and pressure. The key step in constructing the generalized potential is adding to the inexact differential form that gives the heat or work exchanged another differential term which (a) is zero on the paths satisfying the given constraints, and (b) makes the inexact differential into an exact differential. This construction is well known in many contexts as a Legendre or Legendre–Cartan transformation. In Sect. 14.2 we shall examine how such constructions can be carried out, first for reversible processes and then for more general cases. From there, we go on to construct some examples and then see how we can find actual ways to best approach the limits they set.

14.2 Constructing Generalized Potentials

Suppose we want to construct a potential for work for a process with some general constraint that a function $g(P,\ V) = \text{const.}$ or $\mathrm{d}g = 0$. The work itself is the value of the integral of $P\mathrm{d}V$ along the chosen path. We want then to construct an exact differential Ω by adding to $P\mathrm{d}V$ a term $f\mathrm{d}g$ that makes $\mathrm{d}\Omega = P\mathrm{d}V + f\mathrm{d}g$ into an exact differential. That is, we want to make an exact differential of

$$P\mathrm{d}V + f\mathrm{d}g = P\mathrm{d}V + f\left(\frac{\partial}{\partial P}\mathrm{d}P + \frac{\partial g}{\partial V}\mathrm{d}V\right)$$
$$= \left(P + f\frac{\partial g}{\partial V}\right)\mathrm{d}V + \left(f\frac{\partial g}{\partial P}\right)\mathrm{d}P. \tag{14.1}$$

which we can do by making the cross-derivatives equal:

$$\frac{\partial\left(P + f\partial g/\partial V\right)}{\partial P} = 1 + \frac{\partial f}{\partial P}\frac{\partial g}{\partial V} + f\frac{\partial^2 g}{\partial V \partial P} = \frac{\partial\left(f\partial g/\partial P\right)}{\partial V}$$
$$= \frac{\partial f}{\partial V}\frac{\partial g}{\partial P} + f\frac{\partial^2 g}{\partial V \partial P}. \tag{14.2}$$

This rearranges to give the condition

$$\left(\frac{\partial f}{\partial V}\right)_P \left(\frac{\partial g}{\partial P}\right)_V - \left(\frac{\partial f}{\partial P}\right)_V \left(\frac{\partial g}{\partial V}\right)_P \equiv \{f, g\}_{P,V} = 1. \qquad (14.3)$$

The symbol $\{\dots\}$ indicates a Poisson bracket. Thus the new function based on a function f that solves (14.3) makes the expression (14.1) into an exact differential, so that Ω becomes a potential. The solution of (14.3) is, in general, not unique, but all solutions for a given constraint $\mathrm{d}g = 0$ yield the same changes of the potential Ω between the given end points.

This construction is applicable not only for reversible processes but, far more generally, for *quasistatic* processes, that is, for processes that can be described as a series of steps taking the system through a succession of states characterizable by equilibrium values of the conventional thermodynamic variables. These may correspond just to discrete points along a process pathway; it is not necessary that the thermodynamic variables be well defined at all points along the path. This means that we can describe the process we study as a time-parametrized sequence of internally equilibrated states. Then one can prove that there is a function of state Ω whose change $\Delta\Omega$ gives uniquely the value of the maximum work W or heat Q transferred in the given process [3]. One can extend this to processes in which subsystems within the larger system go through their own quasistatic succession of states, but are not in equilibrium with each other.

Thus, the potential can reflect the time dependence of the process explicitly. Suppose that P and V are functions of time t, as $P(t)$ and $V(t)$, respectively, governed by the differential equations $P(t) = F(P, V, t)$ and $V(t) = G(P, V, t)$. Then we can construct invariant differential forms

$$\mathrm{d}\theta_1 = \mathrm{d}P - F\mathrm{d}t \quad \text{and} \quad \mathrm{d}\theta_2 = \mathrm{d}V - G\mathrm{d}t$$

which we can then use to create the exact differential, that of the desired potential,

$$\mathrm{d}P = \mathrm{d}W + f_1\mathrm{d}\theta_1 + f_2\mathrm{d}\theta_2. \qquad (14.4)$$

This is a potential for any process for which $\mathrm{d}\theta_1/\mathrm{d}t = 0$ and $\mathrm{d}\theta_2/\mathrm{d}t = 0$. This can be extended to more variables as well.

14.3 Examples: Systems with Finite Rates of Heat Exchange

One rather simple but straightforward illustrative example is a system that does work but differs from ideal reversible engines by exchanging heat with its reservoirs at finite rates, given by a linear relation,

$$q = \frac{\mathrm{d}Q}{\mathrm{d}t} = \kappa\,(T^{\mathrm{ex}} - T) \text{ or, in terms of entropy change,}$$

$$\frac{\mathrm{d}S}{\mathrm{d}t} = \frac{\dot{Q}}{T} = \frac{\kappa\,(T^{\mathrm{ex}} - T)}{T} \tag{14.5}$$

where κ is the heat conductance and T^{ex} is the reservoir temperature [4]. While the second equation is universally correct for reversible processes, it also can apply to quasistatic processes, and to those such as described in (13.4)–(13.6) of Chap. 13. This is a useful model both because it can be completely solved, and because it has wide applicability. For such a system, minimizing the loss of availability, i.e., of the total work available, is equivalent to minimizing the total entropy production, which, for this system, is achieved by holding the rate of entropy production constant on each branch of the cycle. From this, one can obtain the value of the maximum work that such an engine can provide, a value considerably more realistic than that of the ideal reversible model.

In carrying out this analysis, it is often convenient to choose as the control variable the system's temperature T. One might alternatively choose the reservoir temperature T^{ex} but the analysis is easier if we use T, and it can be controlled in practice by making small adiabatic adjustments along the path of the process. We assume that such adjustments require only infinitesimal times.

The analysis begins with a basic theorem: For this system, minimum entropy production implies a constant rate of entropy production on each branch of the cycle. In the limit of a slow process, this rate is the same for all branches of the cycle. A second theorem follows: Suppose the total cycle time is τ, that the entropy produced on the ith branch of the cycle is σ_i, and that κ is the maximum of the heat conductances on all branches. Then a lower bound for the entropy production, per cycle, is

$$\Delta S \geq \left(\sum_{1} |\sigma_i|\right)^2 / \kappa\tau. \tag{14.6}$$

Furthermore the work such a system can produce is also bounded:

$$W \leq W_{\mathrm{rev}} - T_0 \left(\sum_{1} |\sigma_i|\right)^2 / \kappa\tau. \tag{14.7}$$

Here T_0 is the ratio of the loss of availability ΔA, to the entropy production ΔS,

$$T_0 = \Delta A / \Delta S. \tag{14.8}$$

This relation is not restricted just to reversible and quasistatic processes. Here, we only outline the proofs, which can be found in [4]. The entropy production is the objective function to be minimized. The total cycle period τ is divided into segments for each branch; one then minimizes the entropy production for each branch, for which the initial and final states and time are fixed, using whatever control variables one has chosen. Then one optimizes the initial and final states of all the linked branches so that the branches join continuously, again using the control variables to do the joining. Finally, one optimizes the distribution of times on each branch, within the constraint that the total cycle time is τ.

Another illustration shows how the method can be extended to still more realistic systems. Consider a heat engine whose heat transfer rate depends, as in the previous example, on the difference between the temperatures of the reservoir and the system and which also has friction, which causes a loss of work. (Maximum power cycles require two isothermal branches [5].) The heat transfer relation is that of (14.5), and we assume here that the frictional losses depend linearly on the velocity at which the system changes volume, e.g., on the velocity of a piston. Expressed in terms of the system's volume V, this relation is

$$\mathrm{d}W_{\mathrm{friction}} = -\alpha \dot{V} \mathrm{d}V \tag{14.9}$$

but other forms are also appropriate for some situations. Hence the net useful work produced by the system, in undergoing a small expansion $\mathrm{d}V$ is the inexact differential

$$-\mathrm{d}W_{\mathrm{net}} = P\mathrm{d}V - \alpha \dot{V} \mathrm{d}V = P\mathrm{d}V - \alpha \dot{V}^2 \mathrm{d}t. \tag{14.10}$$

We suppose that the heat generated by friction goes entirely into the environment and does not contribute to the entropy of the system.

We need one more condition to specify enough to make this a solvable problem. For this, we choose to define the time dependence of the volume. Let us choose a sinusoidally varying volume,

$$\dot{V} = \frac{\pi(V_{\max} - V_{\min})}{\tau} \sin(2\pi t/\tau),$$

as in a conventional internal combustion engine.

The cyclic, sinusoidal engine with friction and finite heat transfer requires that the pressure and volume be related by a general polytropic connection,

$$P/P_0 = (V_0/V)^{\varsigma},$$

where, if the process is adiabatic, $\varsigma = \gamma \equiv C_P/C_V$. We suppose that the exponent is constant, whatever it may be. If we substitute the sinusoidal time variation of the volume as given previously into the expression for the work produced by the system, we immediately obtain the condition that the cross derivatives of Ω are equal and

$$\mathrm{d}\Omega = P_0 (V_0/V)^{\varsigma} \mathrm{d}V - \alpha \left[\frac{\pi (V_{\max} - V_{\min})}{\tau}\right]^2 \sin^2 (2\pi t/\tau) \mathrm{d}t. \tag{14.11}$$

This integrates immediately so that the time integral from $t = 0$ to $t = t_0$ is

$$\Omega = \frac{P_0 V_0}{1-\varsigma}\left(\frac{V_0}{V}\right)^{\varsigma-1} - \frac{\alpha\pi}{\tau}\left(V_{\max} - V_{\min}\right)^2 \sin^2\left(\pi t_0/\tau\right) \qquad (14.12)$$

which is the potential for this engine.

14.4 Some More Realistic Applications: Improving Energy Efficiency by Optimal Control

Showing that potentials can be defined and constructed for finite-time processes are only the first two of three steps to successful use of the concept. Knowing that the potentials can be constructed is reassuring, even a justification for going further, but it does not do a thing to improve how we use energy. Finding a way to evaluate the potentials brings us closer to using thermodynamics in real systems, by showing quantitatively what the minimum work or heat exchange would have to be, if a specific process were to operate optimally but at a given non-zero rate or in a fixed finite time. The real fruition of this approach comes in determining a pathway to operate a process as nearly as possible to the ideal, *finite-time* limit. How, we must ask, can we design and operate the process so that the actual work done is as close as we can bring it to the limit given by the change in the finite-time thermodynamic work potential?

It was pointed out very early [3] that the natural way to find such optima is through the application of optimal control theory. In fact the first such application was carried out by Rubin [6, 7], specifically to find the pathways and optimal performance so obtained for a cyclic engine of the sort described above, Rubin found the conditions for optimum power and for optimum efficiency, which of course are normally different. It was in these works that he introduced the term 'endoreversible' to describe a process that could have irreversible interactions with its environment but would be describable internally in terms of the thermodynamic variables of a system at equilibrium. An endoreversible system comes to equilibrium internally very rapidly compared, whatever heat or work exchange it incurs with the outside. It was here that one first saw the comparison of the efficiency for maximum power of the Curzon–Ahlborn engine compared graphically with the maximum efficiency, in terms of a curve of power vs. heat flow. Figure 14.1 is an example of this.

Optimal control theory was a useful tool for finding the pathway, in terms of the time dependence of temperature T and volume V, that yields the optimum power or the optimum efficiency for a given fixed cycle time. We shall only outline the method and show some of the results here; the full derivation is available [6].

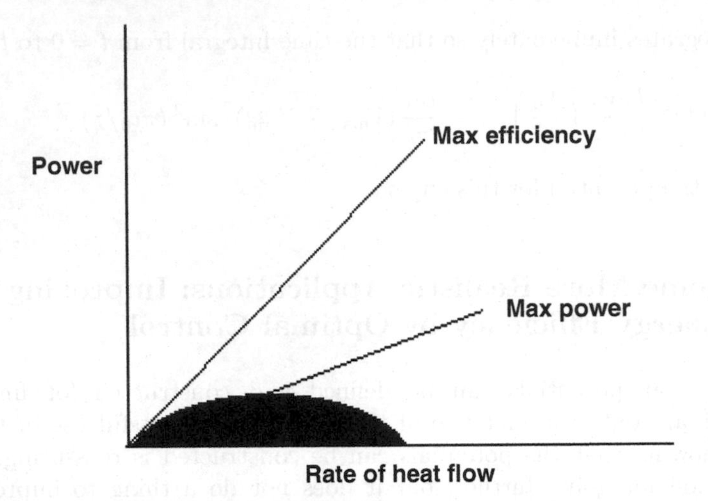

Fig. 14.1. A representation of power delivered vs. heat flow into the system; the two lines correspond to efficiencies. The steeper straight line has the slope of the maximum efficiency, $\eta = 1 - T_L/T_H$ and touches the curve at the point of zero heat flow; the line going through the maximum of the curve crosses that curve at its maximum, and has the slope $\eta = 1 - (T_L/T_H)^{1/2}$

Optimal control theory is a formalism created to be similar in structure to Hamiltonian mechanics. One describes the evolution of a system in terms of some set of state variables and a control function (or functions) which also depends on those variables or related ones. The problem is stated in terms of a goal whose degree of achievement is measured by a performance index or objective function that we can write as a time integral of a time-dependent function L, relating the state variables $\mathbf{x}(t)$ and the control variables $\mathbf{u}(t)$:

$$I = \int_{t_1}^{t_2} L\left[\mathbf{x}\left(t\right), \mathbf{u}\left(t\right)\right] \mathrm{d}t, \tag{14.13}$$

where $I(t_1, t_2)$ is that objective whose value we want to maximize or minimize. The formal procedure treats L analogously to the Lagrangian, from which we define a Hamiltonian $H[\mathbf{x}(t),\ \mathbf{u}(t),\ \psi(t)]$, where $\psi(t)$ plays the role of a Lagrange multiplier so the Hamiltonian can be written

$$H\left(\mathbf{x}, \mathbf{u}, \psi\right) = L\left(\mathbf{x}, \mathbf{u}\right) + \psi \cdot \dot{\mathbf{x}} \tag{14.14}$$

and, just as in the Hamiltonian formalism of mechanics, the time dependences of the adjoint variable ψ and the state variable set $\mathbf{x}$ are simply derivatives of the Hamiltonian with respect to the conjugate variables,

$$\psi(t) = -\partial H/\partial \mathbf{x} \qquad \text{and} \qquad \mathbf{x} = \partial H/\partial \psi. \tag{14.15}$$

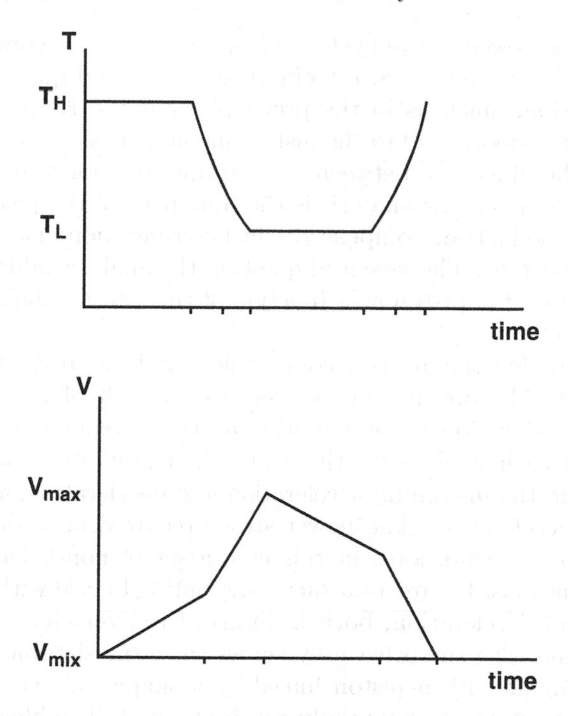

Fig. 14.2. Schematic portrayal of the time history for a Carnot-like engine with friction and finite heat conductance to its reservoirs, optimized to produce maximum average power. The upper figure shows the temperature variation and the lower, the volume changes, as the system goes through its isothermal and adiabatic branches

The desired objective corresponds to the absolute maximum (or minimum) of H over the range of the set of the control variables $\mathbf{u}$.

If one maximizes the mean power produced, the energy per cycle, one obtains an explicit pathway for the cycle, the volume (or piston displacement) and temperature as functions of time. Such an optimal pathway is sketched schematically but in fact from real calculations of the optimal Carnot-like cycle with heat leak and friction, in Fig. 14.2.

14.5 Optimization of a More Realistic System: The Otto Cycle

The Otto cycle is essentially the cycle describing the internal-combustion automobile engine. This is a four-stroke cycle, in contrast to the simpler two-stroke Carnot cycle and the various others, such as the Stirling and Brayton cycles, that operate on a single oscillation of the piston. The Otto cycle consists of an intake expansion, a compression, an expansion resulting from ignition that delivers the power of the cycle, and the final exhaust compression. The model used for this analysis [8, 9] includes heat leak and quadratic, velocity-dependent friction, much as in the previous examples. However the rate of heat loss here is proportional to the instantaneous surface area of the cylinder as well as to the difference between interior and exterior temperatures. The one control variable in this analysis is the time path of the piston. The compression ratio, the fuel/air composition, fuel consumption and total period of the cycle are constant. The essential question the analysis addresses is, 'How can one best move the piston as a function of time, to maximize the average power delivered?'

The solution for the intake, compression and exhaust strokes is very straightforward. The optimal piston velocity in each of these is constant, with a brief acceleration or deceleration at the maximum allowed rate at the juncture of each stroke with the next. The analysis was done both with no constraint on the maximum acceleration and deceleration, and with finite limits on the acceleration. The power stroke required numerical solution of the optimal control equations, in this case a set of non-linear fourth-order differential equations. Figure 14.3 shows the optimal cycle with limits on the acceleration and deceleration, both in terms of the velocity and position as functions of time. The smoother grey curves show the sinusoidal motion of a conventional engine with a piston linked by a simple connecting rod to the drive shaft that rotates at essentially constant speed. The black curves show the optimized pathway.

The most striking characteristic of the optimized time path is the marked deviation from sinusoidal motion in the power stroke, the first stroke shown at the beginning of the cycle. The piston accelerates very rapidly as the fuel–air mixture ignites and then, most important, the piston moves out at essentially its maximum rate. This is precisely what enables this optimized cycle to transform as much of the heat energy as possible into work before it can leak out of the cylinder into the surroundings.

What can such optimization achieve? One criterion is the effectiveness, sometimes called the 'second-law efficiency,' which is the ratio of the work done by the process to that it would do reversibly. The conventional Otto cycle used for this analysis would have an effectiveness of 0.633; the most effective of the optimized engines modeled in the analysis but with a maximum piston velocity of $22.4\,\mathrm{m/s^{-1}}$ would be 0.698; with no velocity constraint, that would go only to 0.705. The model used in the analysis dissipates about 3/5 of its total losses as friction and 2/5 as heat loss. If the engine chosen as the conventional basis for comparison were to lose only about 30% of its total through friction and 70% through heat leak, the effectiveness of the optimized engine would be as much as 17% greater than the conventional engine.

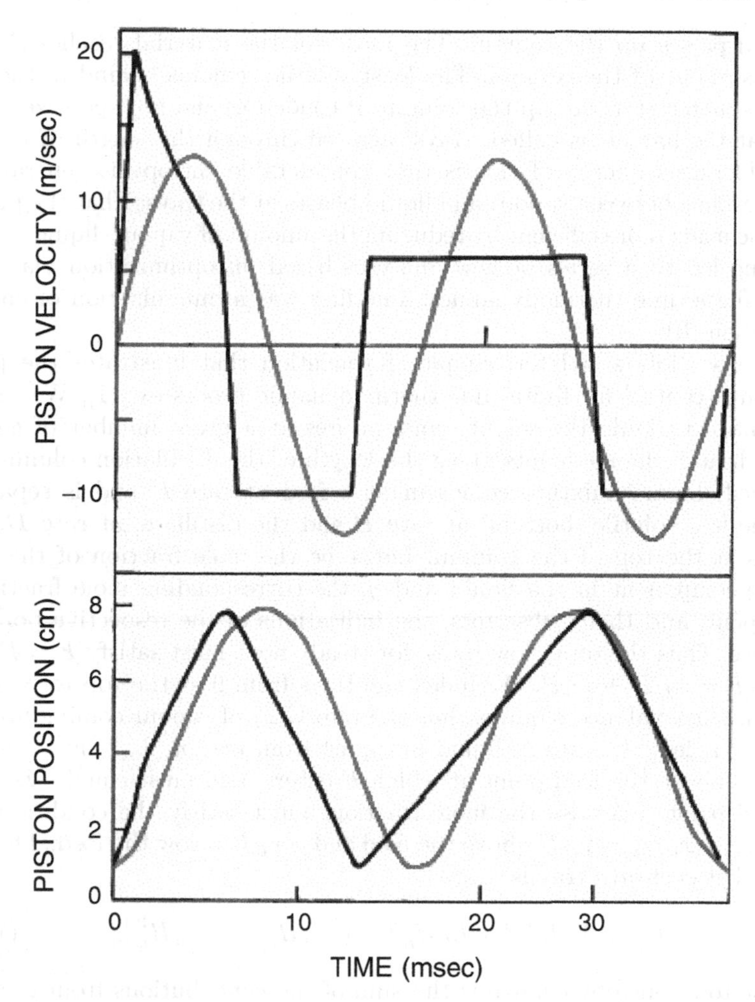

Fig. 14.3. Velocity and position of the piston in a conventional Otto cycle engine (grey curves) and in an engine whose piston path is optimized to give maximum power. The first stroke is of course the power stroke, showing the very rapid acceleration of the piston on ignition and the fast expansion that captures the heat energy

14.6 Another Example: Distillation

One of the most energy-inefficient of widely used industrial processes is distillation, or heat-driven separation processes generally. The traditional distillation column, familiar to students who have had chemistry laboratory courses, has a source of heat at the bottom and a cooling fluid that runs the length of a vertical column, so that there is a temperature gradient, cooling as the material passes up the column. The most volatile material reaches the top and passes out of the column. The least volatile remains behind in the bottom. As material moves up the column, it condenses and re-vapourizes many times at the flat areas called 'trays' arrayed through the length of the column. The consequence of this is that considerable entropy is generated in that recycling between vapour and liquid phases at the trays. That the process could be made more efficient by reducing the amount of vapour–liquid–vapour recycling led to a series of new analyses based on optimization via methods of finite-time thermodynamics. The first was a minimization of entropy production [10].

Here, we follow a later, simpler formulation that illustrates the power of optimal control for finite-time thermodynamic processes [11]. We take as the control variable the set of temperatures at a given number of equally spaced heat-exchange points along the length of the distillation column. The (assumed) binary mixture comes in as a feed at rate F and is separated into the less volatile 'bottom' at rate B and the distillate, at rate D, that collects at the top of the column. Let x be the mole fraction of the more volatile component in the liquid and y, the corresponding mole fraction in the vapour, and their subscripts, the indications of the respective points of reference. Thus the total flow rates, for steady flow, must satisfy $F = D + B$, and $x_F F = x_D D + x_B B$. We index the trays from 0 at the top to N at the bottom. Mass balance requires that the rate V_{n+1} of vapour coming up from tray $n+1$, less the rate of liquid dropping from tray n, L_n, must equal D for trays above the feed point at which F enters, and must equal $-B$ below the feed point. Likewise the mole fractions must satisfy the condition that $y_{n+1}V_{n+1} - x_n L_n = x_D D$ above the feed and $-x_B B$ below the feed. The heat required at each nth tray is

$$Q_n = V_n H_n^{\text{vap}} + L_n H_n^{\text{liq}} - V_{n+1} H_{n+1}^{\text{vap}} - L_n H_{n-1}^{\text{liq}} \qquad (14.16)$$

and the total entropy change is the sum of the contributions from the feed, bottom and distillate,

$$\Delta S^{\text{streams}} = -F s_F + D s_D + B s_B, \qquad (14.17)$$

where the entropies per mole of mass flow, with $i = F,\ D$ or B, are

$$s_i = x_i \left(s_{\text{ref},1} + c_p^{\text{liq},1} \ln \frac{T_i}{T_{\text{ref}}} \right) + (1-x) \left(s_{\text{ref},2} + c_p^{\text{liq},2} \ln \frac{T_i}{T_{\text{ref}}} \right)$$
$$+ R\left[x_i \ln x_i + (1-x_i) \ln (1-x_i) \right] \qquad (14.18)$$

in terms of reference values of the individual entropies. Then the total entropy change becomes

$$\Delta S^{u,\text{sep}} = \sum_{n=1}^{N} \frac{Q_n}{T_n} + \Delta S^{\text{streams}} = \sum_{n=1}^{N} \frac{Q_n}{T_n} + F\left(d s_D + b s_B - s_F \right)$$

in terms of the constants

$$d \equiv \frac{x_F - x_B}{x_D - x_B}, \qquad b \equiv \frac{x_F - x_D}{x_D - x_D}.$$

The entropy associated with heat conduction in each jth tray,

$$\Delta S_j^{u,\mathrm{hx}} = Q_j \left(\frac{1}{T_j} - \frac{1}{T_j^{\mathrm{ext}}} \right)$$

is also included in the optimization. The analysis was carried out for two models of heat conductance, Fourier transfer in which the rate of heat flow is proportional to the difference of the inverses of internal and external temperature, and Newtonian transfer, in which the rate is simply proportional to the temperature difference. The results are almost identical. One finds the external temperatures at each contact point in terms of the conductivity and the internal temperature, and from that set, one obtains the total entropy change produced in the process. The optimization was carried out by computation to minimize the total entropy production, for the separation of an equimolar mixture of benzene and toluene, with the goal of 95% separation in the distillate and bottom. Three columns, with 25, 45 and 65 trays were evaluated. Figure 14.4 shows schematic diagrams of a traditional column and the column with heat exchanges at distributed points.

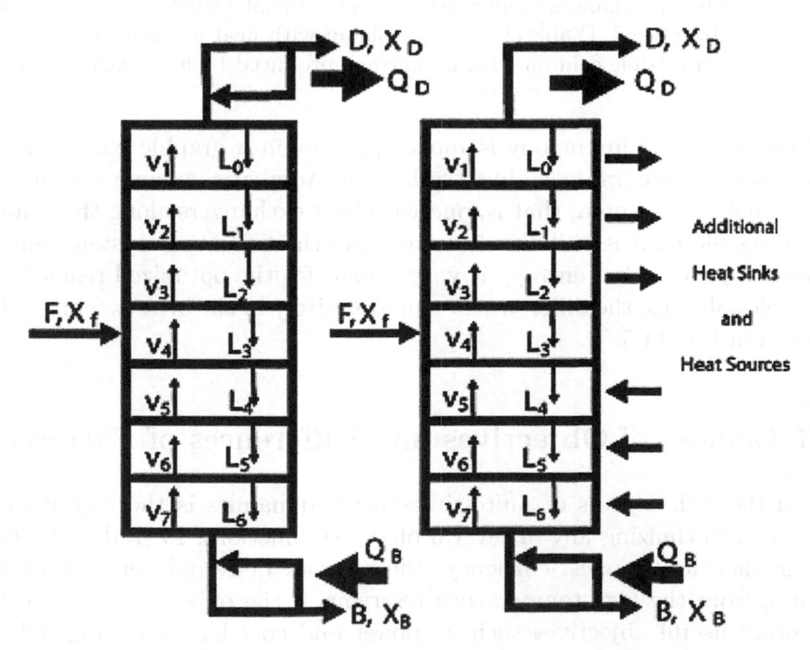

Fig. 14.4. Schematic representations of two distillation columns; *left*, a traditional column; *right*, a column with heat-exchange points along the column

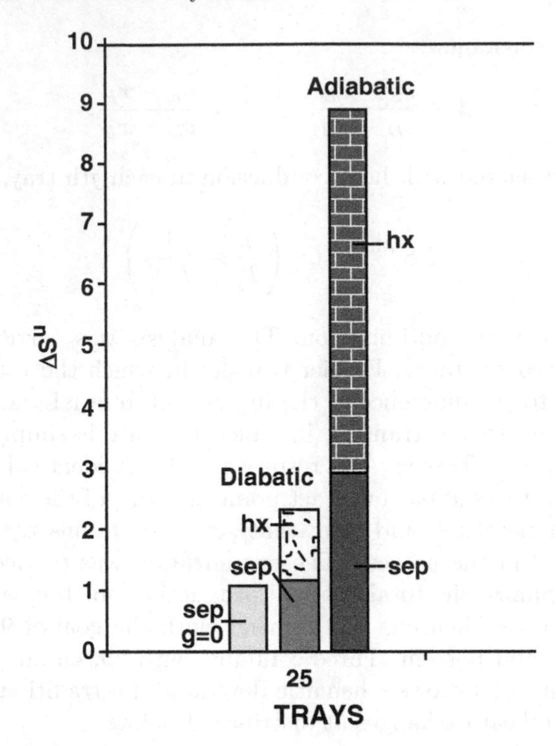

Fig. 14.5. Entropy changes computed for a traditional 'Adiabatic' column of 25 trays, for an optimized 'Diabatic' 25-tray column with heat exchange at each tray, and an ideal, reversible column with no entropy produced by heat exchange, hx

The net saving in entropy is most apparent in a graphic comparison of the entropy change produced in a traditional 'Adiabatic' column and an optimized 'Diabatic' column, that is, one with heat exchangers along the column. The reversible limit is still clearly lower than the finite-time system, but the separation part of that entropy is very similar for the optimized realistic and reversible columns; the difference is almost entirely in the heat exchange. This is shown in Fig. 14.5.

14.7 Choices of Objectives and Differences of Extrema

One of the rich aspects of finite-time thermodynamics is the way it opens options for optimizing any of several objective functions. In traditional equilibrium thermodynamics, efficiency, the network delivered per unit of heat taken in from the high-temperature reservoir, is the only objective one has. The other useful objectives such as power and cost have meaning only for systems operating in finite time. It is useful to get some sense of the differences of operating conditions for a chosen process optimized for different

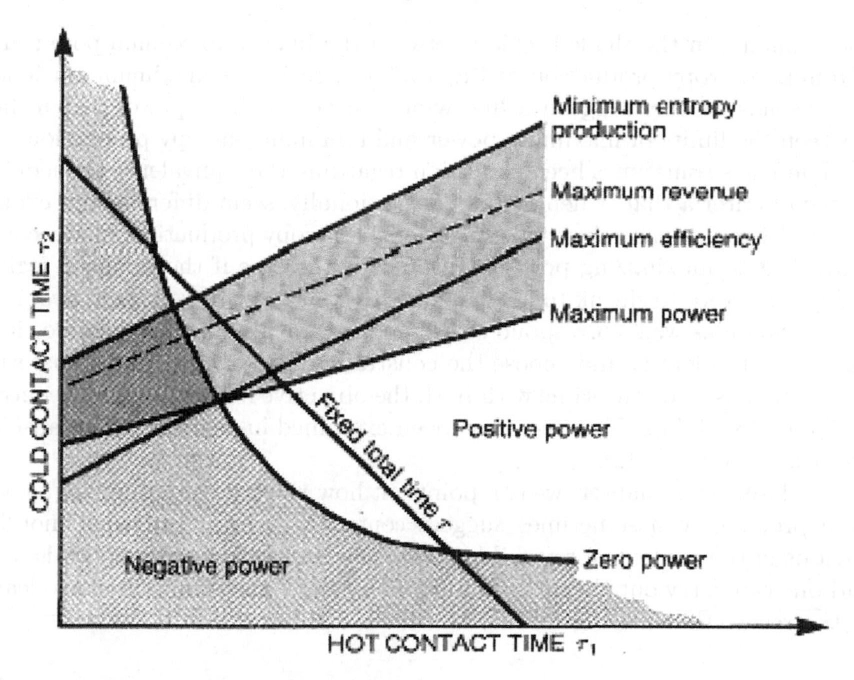

Fig. 14.6. Optima for an endoreversible, finite-time Carnot-type engine with finite heat conductances to its hot and cold reservoirs. The vertical axis represents contact time with the cold reservoir, and the horizontal axis, with the hot reservoir. Conditions for maximum revenue lie in the *shaded region* between the limits of minimum entropy production and maximum power production

objectives. It is illustrative to consider the endoreversible Carnot engine, that is, a Carnot cycle operating endoreversibly but with finite heat conductances to its hot and cold reservoirs. The behaviour of this system can be described graphically in terms of the contact times with those two reservoirs. Any line sloping down at 45° corresponds to operating at a fixed time. For very short contact times, the system can do no network because it cannot absorb or deposit adequate amounts of energy as heat; under those conditions, the engine requires power to operate its cycle. The conditions dividing the positive power delivery regime from the negative define a smooth curve in the space of the two contact times, as shown in Fig. 14.6. The conditions for minimum entropy production and for maximum power define two boundaries, between which the system can operate to maximize revenue. That is, the cost of operating can be expressed in terms of the entropy produced, per cycle, and the return of profit can be expressed in terms of the power delivered. The greater the return for power, relative to the cost of running the engine and generating entropy, the closer one wants to operate near the limit of maximum power. If, on the other hand, the operating costs are high, one wants to operate near the boundary set by minimum entropy production. Hence the profitable range of operation lies in the shaded region between the lines of maximum power and minimum entropy production of Fig. 14.6. Operating at maximum efficiency corresponds to operating on a line which, in the positive power region, lies between the limits of maximum power and minimum entropy production.

There has sometimes been confusion regarding the equivalence or inequivalence of criteria that, when stated conventionally, seem different. For example, under some constraints, minimization of entropy production may become equivalent to maximizing power, as would be the case if the coloured region of Fig. 14.6 were to shrink to a single line. However in general, these two, and other criteria as well, correspond to different operating conditions. Hence it is important to identify and choose the constraints that one's process must satisfy, as well as to do the same with both the objective function and the control variable or variables. This issue has been examined in some detail because of just that confusion [12].

In closing this chapter, we can point out how thinking of optimizing finite-time processes can sometimes suggest control variables that would not be obvious in the context of reversible thermodynamics. The piston path is one, and one can carry out the same kind of analysis for the diesel cycle. We leave to the reader the challenge of finding others.

References

1. F.L. Curzon, B. Ahlborn, Am. J. Phys. **43**, 22–24 (1975)
2. B. Andresen, S.R. Berry, A. Nitzan, P. Salamon, Phys. Rev. A **15**, 2086–2093 (1977)
3. P. Salamon, B. Andresen, R.S. Berry, Phys. Rev. A **15**, 2094–2102 (1977)
4. P. Salamon, A. Nitza, B. Andresen, S.R. Berry, Phys. Rev. A **21**, 2115–2129 (1980)
5. D. Gutkowitz-Krusin, I. Procaccia, J. Ross, J. Chem. Phys. **69**, 3898–3906 (1978)
6. M.H. Rubin, Phys. Rev. A **19**, 1277–1289 (1979)
7. M.H. Rubin, Phys. Rev. A **19**, 1272–1276 (1979)
8. M. Mozurkewich, R.S. Berry, J. Appl. Phys. **53**, 34–42 (1982)
9. M. Mozurkewich, R.S. Berry, J. Appl. Phys. **54**, 3651–3661 (1983)
10. O. Mullins, R.S. Berry, J. Phys. Chem. **88**, 723–728 (1984)
11. M. Schaller, K.H. Hoffman, R. Rivero, B. Andresen, P. Salamon, J. Nonequil. Thermo. **27**, 257–269 (2002)
12. P. Salamon, P. Hoffman, K.H. Schubert S. Berry, R. Andresen, J. Nonequilib. Thermo. **26**, 73–83 (2001)

15

Reduction of Dissipation in Heat Engines by Periodic Changes of External Constraints

15.1 Introduction

In Chaps. 13 and 14 we presented a discussion of heat engines operating in finite time with some irreversible (spontaneous) process, such as heat conduction, contributing to a non-zero power output and dissipation.

In this chapter we continue the analysis of heat engines driven by chemical reactions by investigating the effects of periodic changes of some external constraints on the efficiency of the engine and on the dissipation. The effects of such periodic changes on the efficiency and dissipation of non-linear reactions, including oscillatory reactions, will be taken up in the following chapters.

15.2 A Simple Example

Consider a thermal engine driven by a chemical reaction: Reactants flow into a continuous flow stirred reactor tank (CSTR) which is in contact with a heat exchange fluid at T_{ex}, see Fig. 15.1, taken from [1]. Work may be done by the movable piston in the tank. Exo- or endothermic reactions occur in the tank which change the temperature of the outflowing chemicals from that of the incoming chemicals, assumed to be T_{ex}. The temperature difference between the products and the heat-exchange fluid can be used to run a heat engine. We apply a periodic variation in the external pressure and let the volume of the CSTR vary. We investigate the possibility of producing a positive power output in the pressure–volume work reservoir such that the total power output is larger than that of the heat engine alone. We assume that the reactants are in the gas phase, are perfect gases, have constant heat capacity, and flow into the reactor at a constant rate.

We write the stoichiometric equation for the jth reaction as

$$\sum_i v_{ij} A_i = 0 \tag{15.1}$$

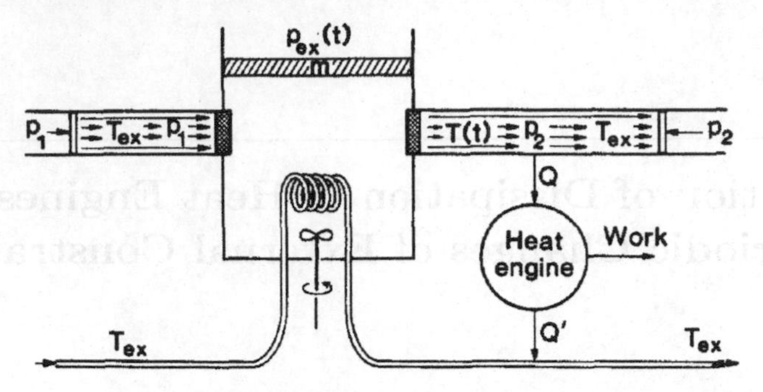

Fig. 15.1. The engine consists of a reactor tank with a movable piston which may do work against an external pressure $p_{\mathrm{ex}}(t)$ and a heat engine which uses the different temperature of products and heat-exchange fluid to produce work

with the stoichiometric coefficient ν_{ij} and the species A_i; the reaction rate (number of reactions per unit time in a unit volume) is r_j. The dynamical equations of motion are

$$
\begin{aligned}
\dot{U} &= j_Q - p\dot{V} + \sum_i \dot{n}_i^{(\mathrm{in})} + h_i(T_{\mathrm{ex}}) - \sum_i \dot{n}_i^{(\mathrm{out})} h_i(T) \quad &\text{(First law)},\\
\dot{V} &= \frac{A^2}{m}(p - p_{\mathrm{ex}}) \quad &\text{(Newton's law)},\\
\dot{n}_i &= \sum_j v_{ij} r_j V + \dot{n}_i^{(\mathrm{in})} - \dot{n}_i^{(\mathrm{out})} \quad &\text{(Reaction kinetics)},
\end{aligned}
\tag{15.2}
$$

where the quantities U, p, T, V are energy, pressure, volume and temperature respectively;

$$
n_i,\ \dot{n}_i^{(\mathrm{in})},\ \dot{n}_i^{(\mathrm{out})}
$$

are, respectively, the number of moles of species i in the reactor, and the number of moles of that species entering and leaving the reactor per unit time; A and m are the area and the mass of the piston. The constituent equations are

$$
pV = \sum_i n_i RT, \tag{15.3}
$$

for ideal gases,

$$
U = U_0 + \sum_i n_i c_{vi} T \tag{15.4}
$$

for energy, and

$$
j_Q = a(T_{\mathrm{ex}} - T) \tag{15.5}
$$

for Fourier's law of conduction of heat. We take the system to be in stable stationary state and when we apply the external variation of pressure the response of the system to that external variation is also periodic. When the system thus responds we take the average of the first law, the first of (15.2), with the definition of the average

$$[\langle\langle\,\rangle\rangle = (1/r)\int_0^\tau \mathrm{d}t\ldots] \tag{15.6}$$

and we obtain

$$\begin{aligned}\langle\dot{U}\rangle = 0 = \langle j_Q\rangle - \langle p\dot{V}\rangle \\ + \left\langle \sum_i \dot{n}_i^{(\mathrm{in})} h_i(T_{\mathrm{ex}}) - \sum_i \dot{n}_i^{(\mathrm{out})} h_i(T) \right\rangle .\end{aligned} \tag{15.7}$$

We can use the average of the third line of (15.2) to eliminate $\left\langle \dot{n}_i^{(\mathrm{in})} \right\rangle$

$$\langle\dot{n}_i^{(\mathrm{in})}\rangle = \langle\dot{n}_j^{(\mathrm{out})}\rangle - \sum_j \nu_{ij}\langle r_j V\rangle \tag{15.8}$$

and obtain for the average power output via the piston

$$\mathcal{P}_p = \langle p\dot{V}\rangle \tag{15.9}$$

$$\begin{aligned}\mathcal{P}_p = \langle j_Q\rangle - \sum_j \langle r_j V\rangle \sum_i v_{ij} h_i(T_{\mathrm{ex}}) \\ + \sum_i \langle \dot{n}_i^{(\mathrm{out})}[h_i(T_{\mathrm{ex}}) - h_i(T)]\rangle .\end{aligned} \tag{15.10}$$

With the enthalpy change of the reaction

$$\Delta H_j(T_{\mathrm{ex}}) = \sum_i v_{ij} h_i(T_{\mathrm{ex}}) \tag{15.11}$$

the enthalpy for ideal gases

$$h_i(T) = h_i^0 + c_{pi}T \tag{15.12}$$

and Fourier's law of heat conduction, (15.5), we obtain

$$\begin{aligned}\mathcal{P}_p = \alpha\langle T_{\mathrm{ex}} - T\rangle - \sum_j \langle r_j V\rangle \Delta H_j(T_{\mathrm{ex}}) \\ + \sum_i c_{pi}\langle \dot{n}_i^{(\mathrm{out})}(T_{\mathrm{ex}} - T)\rangle .\end{aligned} \tag{15.13}$$

Next we calculate the power output of the Carnot engine operating between the temperature of the products, T, and the temperature of the heat bath T_{ex}. We should run the engine as we considered in Chap. 13, that is irreversible.

However for comparing the power output of the engine with (mode b), and without (mode a), an external variation of the pressure we shall assume for simplicity that the heat engine is ideal, that is reversible, with the rate of the work output determined by the rate of product outflow. The heat per unit time that is transferred to or from the products is

$$\sum_i \dot{n}_i^{(\mathrm{out})} c_{pi}(T - T_{\mathrm{ex}}). \tag{15.14}$$

The heat capacity at constant pressure appears in that equation because we take the products to be at the constant pressure p_2. The power produced by the Carnot engine, P_{c}, is then [2]

$$\mathcal{P}_{\mathrm{c}} = \sum_i c_{pi} \left\langle \dot{n}_i^{(\mathrm{out})} \left(T - T_{\mathrm{ex}} + T_{\mathrm{ex}} \ln \frac{T_{\mathrm{ex}}}{T} \right) \right\rangle, \tag{15.15}$$

which leads to the total power output

$$\begin{aligned} \mathcal{P} &= \mathcal{P}_{\mathrm{p}} + \mathcal{P}_{\mathrm{c}} = \alpha \langle T_{\mathrm{ex}} - T \rangle - \sum_j \langle r_j V \rangle \Delta H_j(T_{\mathrm{ex}}) \\ &+ \sum_i c_{pi} T_{\mathrm{ex}} \left\langle \dot{n}_i^{(\mathrm{out})} \ln \frac{T_{\mathrm{ex}}}{T} \right\rangle. \end{aligned} \tag{15.16}$$

The process of pushing the gases through the reactor requires no work because we neglect viscous effects; the products in their final state have the same temperature as the reactants and there is no change in the number of moles during the reaction.

The various expressions for the power output can be simplified if we consider only one reaction

$$\sum_i v_i A_i = 0 \tag{15.17}$$

and all species are withdrawn from the reactor at the same rate. With these assumptions the average conversion rate of chemicals $\langle rV \rangle$ is the same in mode (a) and (b). We denote with $\langle\rangle_a$ and $\langle\rangle_b$ the time averages in the respective modes, and write the power output via the piston (see (15.13))

$$\mathcal{P}_{\mathrm{p}}^{(b)} = \left(\alpha + \sum_i c_{pi} \dot{n}_i^{(\mathrm{out})} \right) (\langle T \rangle_a - \langle T \rangle_b). \tag{15.18}$$

A positive power output via the piston implies that the average temperature of the reactor contents in the perturbed case is lower than in the unperturbed case. The difference in total power output of the perturbed and unperturbed case is

$$\Delta \mathcal{P} = \mathcal{P}^{(b)} - \mathcal{P}^{(a)}, \tag{15.19}$$

which is

$$\Delta\mathcal{P} = \alpha(\langle T\rangle_a - \langle T\rangle_b) + \sum_i c_{pi}\dot{n}_i^{(\text{out})} T_{\text{ex}}(\langle \ln T\rangle_a - \langle \ln T\rangle_b). \tag{15.20}$$

If the system is in a stable stationary state then it follows that a positive power output via the piston in case (b) also produces an increased total power output.

From conservation of energy we have the condition that if $\mathcal{P}^{(b)} > \mathcal{P}^{(a)}$ then the total work in mode (b) is larger than that in mode (a), that is $|W^{(b)} - W^{(b)}| = |Q^{(b)} - Q^{(a)}| > 0$. Thus in the perturbed mode a larger amount of heat is converted into work for the same average reaction rate, see Fig. 15.2, taken from [1],

The difference in the global entropy change per mole of chemical throughput in the mode (b) minus that of mode (a) is

$$-\frac{Q^{(b)} - Q^{(a)}}{T_{\text{ex}}}. \tag{15.21}$$

Thus if $\Delta\mathcal{P} > 0$ then mode (b) has a smaller increase in entropy, a smaller increase in entropy production, and a larger efficiency.

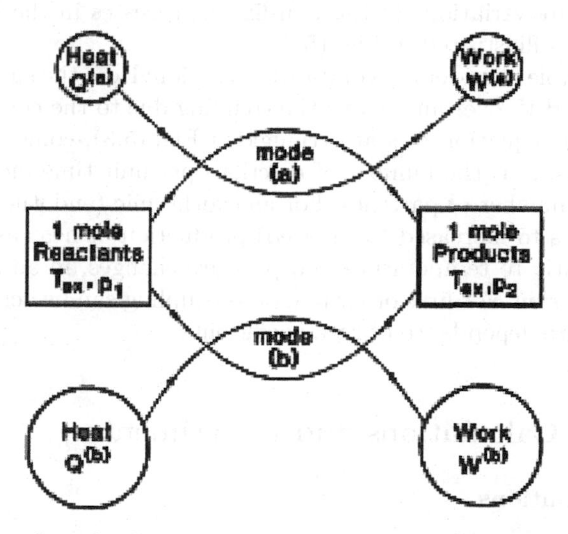

Fig. 15.2. For each mole of reactants converted into products, a certain amount of heat is converted into work. More heat is converted into work in the system which is subject to external pressure variations and which produces pressure–volume work [mode (b)]. The figure is drawn for an endothermic reaction for which $Q^{(b)}$, the net quantity of heat flowing from the reservoir to the system (reactor tank and heat engine) is larger than $Q^{(a)}$

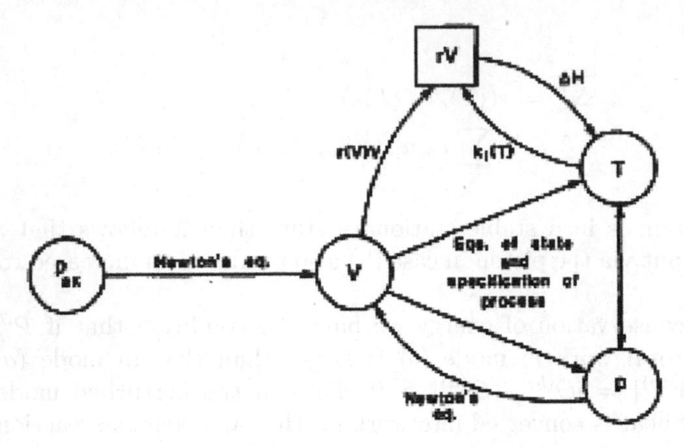

Fig. 15.3. The reaction rate r opens a new channel among the physical properties p, V and T of the system which can change the phase relations among those quantities such that the average pressure on expansion is larger than on compression. From [3]

A positive power output in the work reservoir necessitates that the pressure on expansion be larger, on the average, than on compression. This requirement can be met by control of the relative phase of the external pressure and the volume of the system; the control is obtained by a suitable coupling of the external pressure variations to the non-linear processes in the reaction tank. This coupling is illustrated in Fig. 15.3.

As an example consider a second-order reaction: Here the suitable coupling between p_{ex} and V, over and above the coupling due to the compressibility of the system (the equation of state triangle in Fig. 15.3), comes from the fact that on decreasing V the number of reactions per unit time increases as V^{-1} for fixed total number of particles. For an exothermic (endothermic) reaction this effect leads to increased (decreased) production of heat as V decreases, and consequently to temperature and pressure changes, in addition to those due to compression. For first-order reactions coupling can be achieved through the temperature dependence of rate coefficients.

15.3 Some Calculations and Experiments

15.3.1 Calculations

We turn now to some calculations and experiments to substantiate all the points made so far. In [4] calculations are reported for the combustion reaction

$$CH_4 + 2O_2 \rightarrow CO_2 + 2H_2O. \tag{15.22}$$

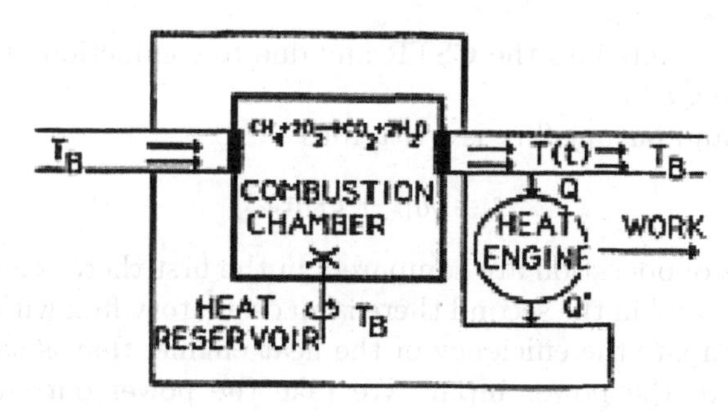

Fig. 15.4. Combustion chamber (CSTR) stirred and coupled to a thermal engine. From [4]

The calculations were made to simulate an experiment in an apparatus shown schematically in Fig. 15.4, which is similar to that shown in Fig. 15.1, in which the temperature of the products T is measured.

The reactants enter the CSTR at the bath temperature, react in the CSTR and raise the temperature. The thermal engine is supposed to operate as an ideal Carnot engine, see [2]. The stoichiometric ratio of oxygen to methane is taken to be 2:1 and the evolution equations [1] of the temperature and concentrations are

$$\begin{aligned}
&C\left(\frac{\mathrm{d}T}{\mathrm{d}t}\right) = -\Delta H(T)r + (\alpha + 3jc_p c_0)(T_B - T),\\
&\frac{\mathrm{d}\,[\mathrm{CH_4}]}{\mathrm{d}t} = -r + j(c_0 - [\mathrm{CH_4}]),\\
&[\mathrm{O_2}] = 2[\mathrm{CH_4}],\\
&C = 3c_{\Delta c_v}c_0 - 3\Delta c_v(c_0 - [\mathrm{CH_4}]),\\
&c_p = (c_{p,\mathrm{CH_4}} + 2c_{p,\mathrm{O_2}})/3,\\
&c_v = (c_{v,\mathrm{CH_4}} + 2c_{p,\mathrm{O_2}})/3,\\
&\Delta c_v = c_v - (c_{v,\mathrm{CO_2}} + 2c_{v,\mathrm{H_2O}})/3,
\end{aligned} \tag{15.23}$$

where c_{pi} and c_{vi} are heat capacities at constant pressure and volume, respectively, of the chemical species i, α is a heat transfer coefficient multiplied by the area of the transfer surfaces and divided by volume of the CSTR; c_0 is the concentration of methane in the input stream; ΔH is the enthalpy change of the stoichiometric reaction, (15.22), which we approximate by

$$\Delta H(T) = \Delta H_0(T_B) + \sum_i \nu_i c_{pi}(T - T_B), \tag{15.24}$$

where ν_i are the stoichiometric coefficients in (15.22). The term j in the first two lines of (15.23) gives changes in concentrations due to input and output fluxes. The first term in the first line of (15.23) is the heat generated by the reaction in unit time, and the second term describes energy changes due to the fluxes of reactants into the CSTR and due to conduction of heat through the walls of the CSTR.

The input and output flux is written as

$$j = j_0[1 + \epsilon \sin \omega t] \tag{15.25}$$

and two modes of operation are compared: in the first there is a constant flux, that is ε is zero, and in the second there is an oscillatory flux with amplitude ε. We wish to compare the efficiency of the heat engine, that is the ratio of the power output to the power input. We take the power output of the ideal Carnot engine (see [2]) to be proportional to the the influx j

$$\begin{aligned} P_0 = 3V_j\{c_p c_0 - \Delta c_p(c_0 - [\mathrm{CH_4}])\} \\ \cdot [\mathrm{T} - T_B(1 + \ln(T/T_B))], \end{aligned} \tag{15.26}$$

and the efficiency to be

$$\eta = (P_0)/\langle P_i \rangle, \tag{15.27}$$

where P_i is the power input. In (15.26) V is the volume of the combustion chamber and

$$\Delta c_p = c_p - (c_{p,\mathrm{CO_2}} + c_{p,\mathrm{H_2O}})/3. \tag{15.28}$$

There are at least two choices for the formulation of the power input, depending on whether all the fuel in the input is counted in the cost of the power input or only the fuel that is burned. In the first choice we have for the efficiency

$$\eta_1 = \langle P_0 \rangle / \langle V T_B 3 j c_p c_0 \rangle, \tag{15.29}$$

in which the heat content of the input flux is approximated by the denominator.

These equations were solved numerically and in Fig. 15.5 we show the ratio of the efficiency as defined in (15.29) for an oscillatory influx of reactants to that efficiency for a steady (constant) influx vs. ω, the frequency of the oscillatory influx, (15.25). The autonomous system, with a steady influx, is in a stable focus, that is the autonomous system on being perturbed briefly returns to the stable state with an oscillatory component. In order to emphasize the effects of an oscillatory influx, conditions were chosen for the steady influx such that only 10% of the heat input is converted to work.

We see that substantial increases in efficiency can be achieved with an oscillatory influx, as much as 30% for the conditions chosen. Variations in the ratio of efficiencies shown in Fig. 15.5 depend both on the amplitue and frequency of the periodic perturbation of the input (and output) flux. The increase in the ratio of frequencies at certain frequencies ω are related to resonance phenomena and appropriate phase relations of the response of the system to the oscillatory influx. At higher amplitudes of oscillatory perturbations there are two resonance peaks (an issue we shall re-visit in later chapters in which we dicuss perturbing chemical reactions).

The expression for the power output, (15.26), is the product of the flux j, which is an effective rate coefficient with units inverse time, t^{-1}, and a

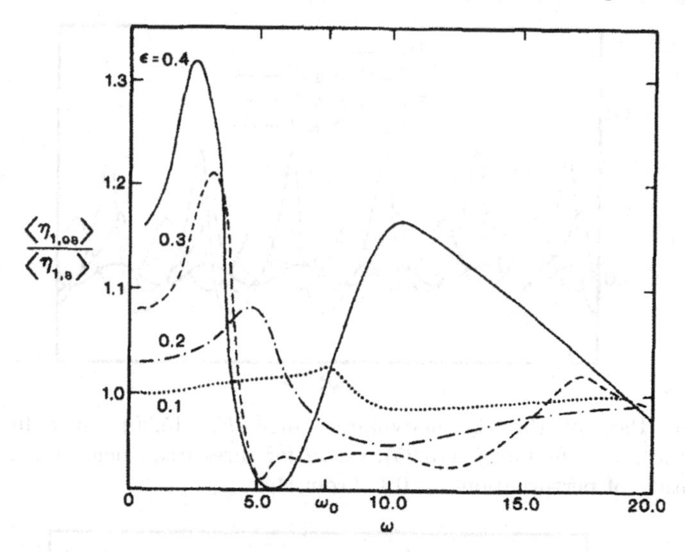

Fig. 15.5. Plot of the ratio of the efficiency, (15.29), for the case of an oscillatory influx into the CSTR, to that for a steady influx, vs. the frequency of the oscillatory influx ω, for four different amplitudes of perturbation ε, (15.25), the smallest being 0.1 (*dotted line*), and the largest 0.4 (*solid line*). The symbol ω_0 denotes the frequency of the damped oscillation in the autonomous system. From [4]

composite term with units of energy (since power equals energy per unit time). The composite term is sometimes referred to as a thermodynamic force

$$F = V[T - T_B(1 + \ln(T/T_B))] \\ \cdot 3 \cdot [c_p c_0 - \Delta c_p(c_0 - [\mathrm{CH_4}])]. \tag{15.30}$$

In Fig. 15.6 we show a plot of the thermodynamic force, (15.30), and the flux, (15.25), vs. ωt, the product of the perturbation frequency and the time, for the case $\varepsilon = 0.1$. The changes in efficiency for that amplitude of perturbation are small, yet much can be learned. At a frequency of $11.5\,\mathrm{s}^{-1}$ the force is out of phase with the influx and the ratio of efficiency dips below unity. The imposition of the perturbation lowers the efficiency in this case. However at the frequency of $7.6\,\mathrm{s}^{-1}$ the flux and force are nearly in phase and the ratio of efficiency has increased to 1.03. Here the perturbation frequency is very close to the frequency of the autonomous system and the resonance nature of the response is apparent.

A plot of the phase relation of the flux and the force for the larger amplitude of perturbation $\varepsilon = 0.4$ is shown in Fig. 15.7. For this case, as seen in Fig. 15.5, the plot of relative efficiency vs. frequency has two maxima. At high frequency of external perturbation, $\omega = 20.0\,\mathrm{s}^{-1}$, the amplitude of the force is small and out of phase with the flux and the ratios of efficiencies are

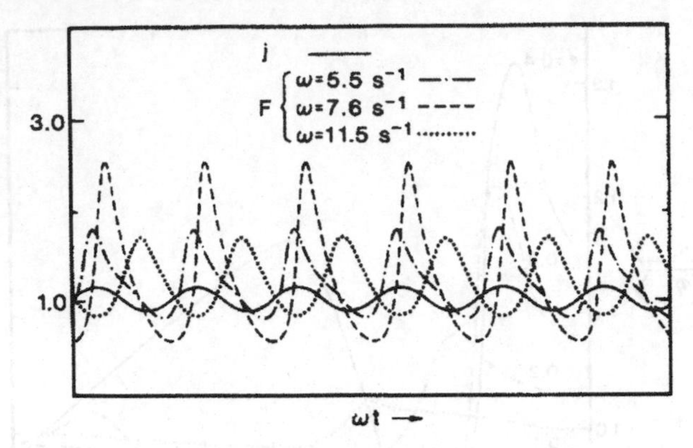

Fig. 15.6. Plot of the thermodynamic force F, (15.30), in reduced units, $F/2.4VRT_{\mathrm{B}}c_0$, and the flux j, (15.25), vs. ωt for three frequencies of perturbation. The amplitude of perturbation $\varepsilon = 0.1$. From [4]

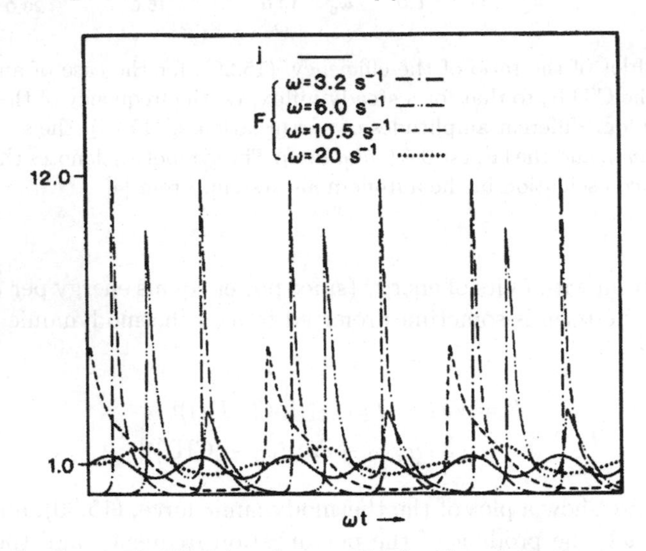

Fig. 15.7. Similar to Fig. 15.6 but for amplitude of perturbation $\varepsilon = 0.4$ and for the frequencies of perturbation as shown in the figure. From [4]

below unity. At the frequency of $10.5\,\mathrm{s}^{-1}$ the amplitude of the force is larger and nearly in phase with the flux; the ratio of efficiencies is about 1.26. At the frequency $\omega = 6.0\,\mathrm{s}^{-1}$ the amplitude is larger but out of phase with the flux and that is close to the minimum between the two maxima in Fig. 15.5. The response for each of the three highest frequencies has a period of twice that of the external perturbation. The period of response changes to that of the perturbation at $\omega = 5.3\,\mathrm{s}^{-1}$. Finally, at a frequency of $3.2\,\mathrm{s}^{-1}$ the amplitude of the force is large, still increasing, and is in phase with the flux; this is close to the first peak in Fig. 15.5.

Further calculations are reported in [5, 6].

15.3.2 Experiments

Experiments were made on the combustion of methane with oxygen [7] in an apparatus somewhat similar to that of Fig. 15.4 and shown in Fig. 15.8.

Measurements consist of the temperature of the products and their chemical composition. Typical results are shown in Fig. 15.9 in plots of an oscillatory influx and the temperature of the reaction products vs. time.

In (a) and (b) there is an efficiency increase of 1.008 and 1.08, respectively, in comparing the oscillatory input flux with constant input flux; the temperature variation and the flux are in phase. In (c) the efficiency is 0.976 and the temperature variation and the flux are not in phase. The changes in efficiency are smaller here than in the calculations, Fig. 15.5, since the possible conversion to work, for constant influx, is more efficient here (about 26%) compared to about 10% in the example in the calculations. Further experiments are

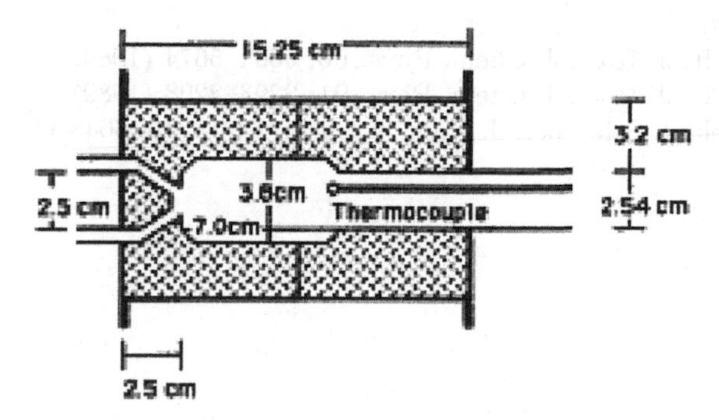

Fig. 15.8. Sketch of the combustion chamber. Taken from [7]

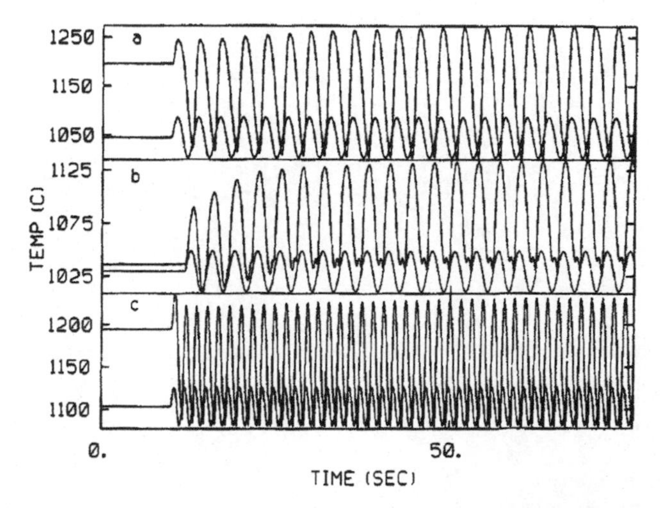

Fig. 15.9. Plots of the temperature of the reaction products (*solid line* full scale) and the input flux (*solid line*, small amplitude) vs. time. See the text. Taken from [7]

reported in [8]. For an analysis of the efficiency of power production in simple non-linear electro-chemical systems see [9].

Acknowledgement. This chapter is based on [1,3,4,7].

References

1. C. Escher, J. Ross, J. Chem. Phys. **1985**, **82**, 2453–2456 (1985)
2. H.B. Callen, *Thermodynamics*, (Wiley, New York, 1960)
3. C. Escher, A. Kloczkowski, J. Ross, J. Chem. Phys. **82**, 2457–2465 (1985)
4. H.M. Schram, M. Schell, J. Ross, Efficiency of a combustion reaction in power production. J. Chem. Phys. **88**, 2730–2743 (1988)
5. A. Hjelmfelt, R. Harding, J. Ross, J. Chem. Phys. **91**, 3677–3684 (1989)
6. A. Hjelmfelt, R.H. Harding, K. Tsujimoto, J. Ross, J. Chem. Phys. **92**, 3559–3568 (1990)
7. A. Hjelmfelt, J. Ross, J. Chem. Phys. **90**, 5664–5674 (1989)
8. A. Hjelmfelt, J. Ross, J. Chem. Phys. **91**, 2293–2298 (1989)
9. A. Hjelmfelt, I. Schreiber, J. Ross, J. Chem. Phys. **95**, 6048–6053 (1991)

16

Dissipation and Efficiency in Biochemical Reactions

16.1 Introduction

After the discussions of dissipation and efficiency in thermal engines we turn to similar considerations for chemical reactions. The main focus will be on oscillatory chemical reactions, as here positive and negative changes in efficiency can be effected in various ways, with illustrations taken frequently from biochemical reactions.

16.2 An Introduction to Oscillatory Reactions

Many chemical and biochemical reactions can be in an oscillatory regime in which the concentrations of intermediates and products vary in a regular oscillatory way in time; the oscillations may be sinusoidal but usually are not. Sustained oscillations require an open system with a continuous influx of reactants; in a closed system oscillations may occur initially when the system is far from equilibrium, but disappear as the system approaches equilibrium. A simple example of an oscillatory reaction is the Selkov model [1]

$$A \underset{k_2}{\overset{k_1}{\rightleftarrows}} S, \qquad S + 2P \underset{k_4}{\overset{k_3}{\rightleftarrows}} 3P, \qquad P \underset{k_6}{\overset{k_5}{\rightleftarrows}} B, \tag{16.1}$$

where A and B are controllable bath concentration and the substrates S and P are the freely responding internal concentrations. If A and B are chosen such that the system is far from equilibrium then oscillations of S and P may occur. It is the non-linearity of the second step in the Selkov model that leads to the possibility of oscillatory concentrations. See [2, 3] for extensive discussions of oscillatory chemical and biochemical reactions. Here we are interested in the dissipation of oscillatory reactions, in the dissipation of externally driven oscillatory reactions, and conversely in the efficiency such processes of converting the Gibbs free energy of the overall reaction to work.

If in such systems there is the possibility of more than one set of products, then the distributions into these several products may be effected by means of external periodic variations of constraints such as influx of reactants or temperature. For a review of earlier work of forced oscillations in systems of chemical interest see [4].

The dissipation, or the rate of entropy production, for chemical reactions is given by

$$\sigma_{\mathrm{R}} = (1/T) \sum A_k v_k, \tag{16.2}$$

where the sum extends over all elementary steps of the reaction mechanism of the reaction, T is the temperature of the isothermal reaction, A_k the affinity (the negative of the Gibbs free energy change) of the kth step, and v_k the rate of that step. If each species is an ideal solute, with standard state of unit molarity, then the affinity can be rewritten as

$$A_k = RT \ln \left(v_k^+ / v_k^- \right), \tag{16.3}$$

where the plus (minus) denotes the rate in the forward (backward) reaction. Hence the dissipation is

$$\sigma_{\mathrm{R}} = \Sigma v_k \ln \left(v_k^+ / v_k^- \right). \tag{16.4}$$

Let us formulate the dissipation for a system with two internal variable (X, Y) coupled to an input bath A and an output bath B. The internal reactant X is converted to the variable Y by an arbitrary non-linear reaction mechanism

$$\dot{Y}^{\mathrm{int}} = -\dot{X}^{\mathrm{int}} = f(X, Y), \tag{16.5}$$

where the superscript 'int' denotes consideration of only internal contributions to the fluxes of X and Y. The input and output baths also couple to the system and hence we have

$$\dot{X} = \dot{X}^{\mathrm{in}} + \dot{X}^{\mathrm{int}}, \qquad \dot{Y} = \dot{Y}^{\mathrm{int}} + \dot{Y}^{\mathrm{out}}. \tag{16.6}$$

For the system–bath exchange we take first-order reactions

$$\dot{X}^{\mathrm{in}} = kA - \bar{k}X, \qquad \dot{Y}^{\mathrm{out}} = k'B - \bar{k}'Y. \tag{16.7}$$

The bath concentrations are externally controlled whereas the concentrations of X and Y are free to respond.

Except for the chemical degree of freedom, we assume that all other equilibrations (translation, rotations, vibrations) take place in times shorter than the chemical reactions; further we assume the solutions are dilute. Hence the chemical potential of a species is

$$\mu_x = \mu_x^0 + k_B T \ln X \tag{16.8}$$

and the dissipation of chemical energy in the reactions of our $(X,\ Y)$ system is

$$D_{AX} = \frac{1}{\tau} \int_0^{\tau} (\mu_A - \mu_X) \dot{X}^{\mathrm{in}} \mathrm{d}t, \qquad \text{(input)}$$

$$D_{XY} = \frac{1}{\tau}\int_0^{\tau} (\mu_X - \mu_Y)\, f(X,Y)\, \mathrm{d}t, \qquad \text{(interior)}$$
$$D_{YB} = \frac{1}{\tau}\int_0^{\tau} (\mu_Y - \mu_B)\left(-\dot{Y}^{\mathrm{out}}\right) \mathrm{d}t. \qquad \text{(output)}$$
$$(16.9)$$

The integrals extend over a full period of an oscillation, τ, for oscillatory reactions. Combining the above equations we find for the total dissipation of this (A, X, Y, B) system to be the sum

$$D_{AX} + D_{XY} + D_{YB} \tag{16.10}$$

that is

$$D = \frac{1}{\tau}\int_0^{\tau} \left(\mu_A \dot{X}^{\mathrm{in}} + \mu_B \dot{Y}^{\mathrm{out}}\right) \mathrm{d}t. \tag{16.11}$$

The dissipation is a production of heat assumed to be conducted away sufficiently fast so that the reaction system is isothermal. Later in the chapter we shall show calculations of the dissipation of a part of glycolysis for stationary states and oscillations.

Next we consider an oscillatory system, such as (16.1), in which there is an oscillatory input of the concentration of the reactant A. In that case the response of the system is not only at the natural frequency of the system but also at other frequencies related to the frequency of the external perturbation of the reactant A. The variety of responses of a typical system are shown in Fig. 16.1.

For a given choice of the perturbation amplitude and ω_{p} the response of the system can be periodic, biperiodic or chaotic. Periodic trajectories (responses) lie in a region called entrainment bands which approach the abscissa

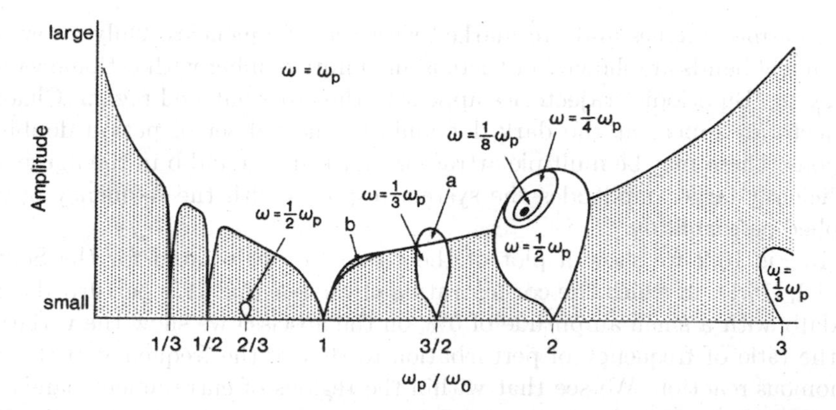

Fig. 16.1. Response of an oscillatory system, a limit cycle, with autonomous frequency ω_0, to a sinusoidal external perturbation of reactant with frequency ω_{p}. Plot of the amplitude of the perturbation vs. ratio of frequencies. For explanatory discussion see text. Taken from [4]

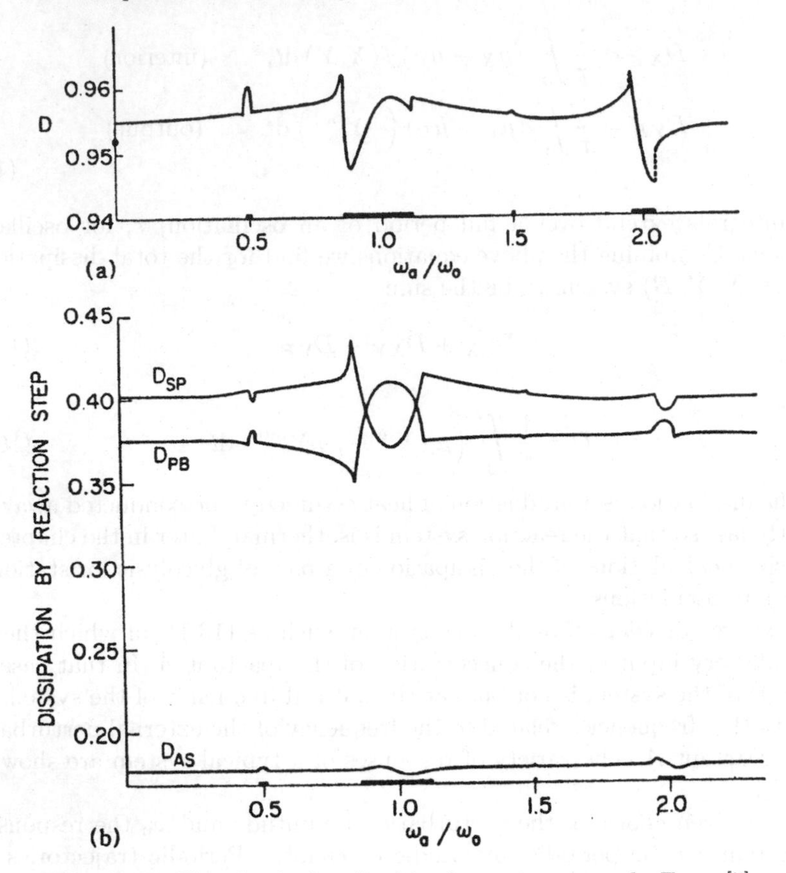

Fig. 16.2. The plot is described in the next paragraph. From [5]

with narrow tongues and are marked with their frequencies. Only a few entrainment bands are shown but there is an infinite number with rational values of $\omega_{\mathrm{p}}/\omega_0$. Biperiodic trajectories appear in the cross-hatched region. Chaotic trajectories appear at the dark dot amidst a nested set of period doubling regions. There may be multiple attractors marked at a and b in the figure. At sufficiently large amplitudes the system responds with the frequency of the applied perturbation.

In Fig. 16.2 we show a plot of the dissipation calculated for the Selkov model, (16.1), in which the concentration of the reactant A is perturbed sinusoidally with a small amplitude of 5%; on the abscissa we show the variation of the ratio of frequency of perturbation to that of the frequency of the autonomous reaction. We see that within the regions of entrainment bands, at 0.5, 1.0, 2.0, and barely visible at 1.5, the dissipation varies significantly. The variations are small in magnitude because of the small amplitude of the perturbation. Again, later in the chapter we show a similar calculation for a part of glycolysis.

16.3 An Oscillatory Reaction with Constant Input of Reactants

Consider the following model for the reaction mechanism of the beginning part of glycolysis (Fig. 16.3) [6] for the production of the energy rich chemical species ATP (adenosine triphosphate), which is the universal energy coinage in biochemical reactions.

In the first reaction glucose reacts with ATP to produce ADP and PEP; the enzyme for this first step is hexokinase; the notation is similar for the remaining reaction steps, with the enzymes as indicated. The rate of influx of glucose into the system is constant. Due to the feedback mechanisms in both the PFK and PK reactions chemical oscillations of some species may occur, see Fig. 16.4. These oscillations have been observed [1] and are also obtained from numerical solutions of the deterministic mass action rate equations of the model in Fig. 16.1 for given glucose inflow conditions, see Fig. 16.4.

The four different periods of oscillations are for different experimental conditions; for a given set of experimental conditions all species oscillate with the same period. For lower glucose inflow oscillations cease and the system is in a stable stationary state (a node or a focus).

In Fig. 16.5 we plot the period of chemical oscillations in the system vs. the total adenine nucleotide concentration, that is the sum of the concentrations of ATP, ADP and AMP, labelled A(MDT)P, for several values of influx conditions. At low values of A(MDT)P there are no oscillations. The oscillations appear at the marginal stability point, and after that the period varies.

Curve c in Figs. 16.5–16.7 is closest to physiological conditions. The stationary concentration of ATP is constant for the calculations on the mechanism in Fig. 16.3 and hence a decrease of A(MDT)P, going to the right in

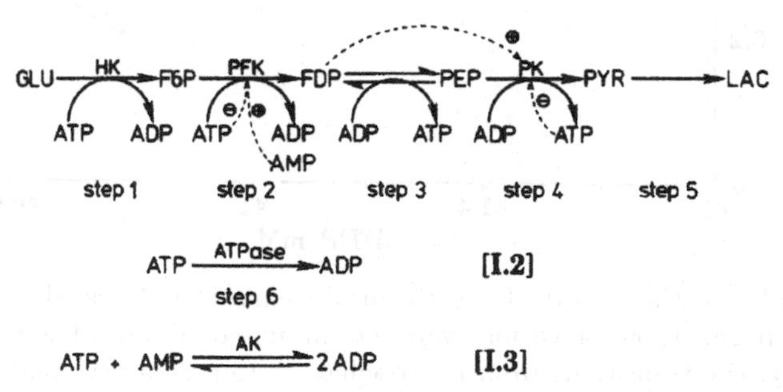

Fig. 16.3. Model for glycolysis. *Arrows* in one direction indicate almost irreversible reactions. *Arrows* in both directions indicate reactions almost at equilibrium. *Broken lines* indicate activation (with encircled plus sign) or inhibition (with encircled minus sign) by metabolites. GLU is glucose, HK is hexokinase, F6P is fructose 6 phosphate, PFK is phosphofructose kinase, FDP is fructose 1,6 bi-phosphate, PEP is phosphoenolpyruvate, PK is pyruvate kinase, PYR is pyruvate, and AK is adenylate kinase and LAC is lactate. From [7]

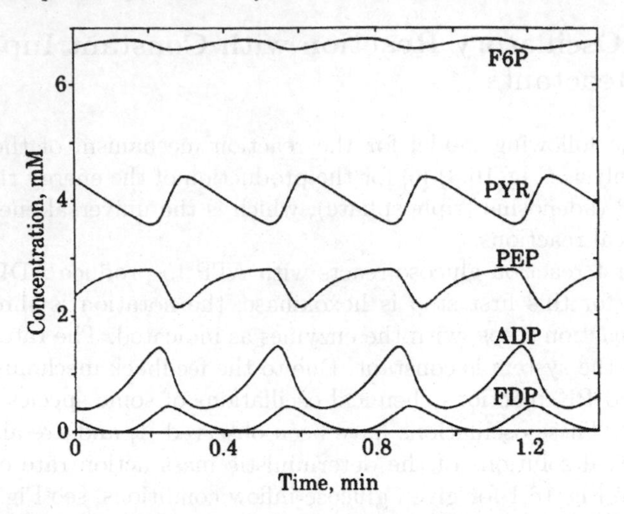

Fig. 16.4. Time dependence of the concentrations of Fru-6-P, Fru-1,6-P_2, P-e-Prv, Prv and ADP as calculated from the model. The of Glc and LAC were assumed to be constant. The figure shows four different oscillation periods. The concentration of Fru-6-P has been reduced by a factor of four. Abbreviations are given in the caption of Fig. 16.3. From [7]

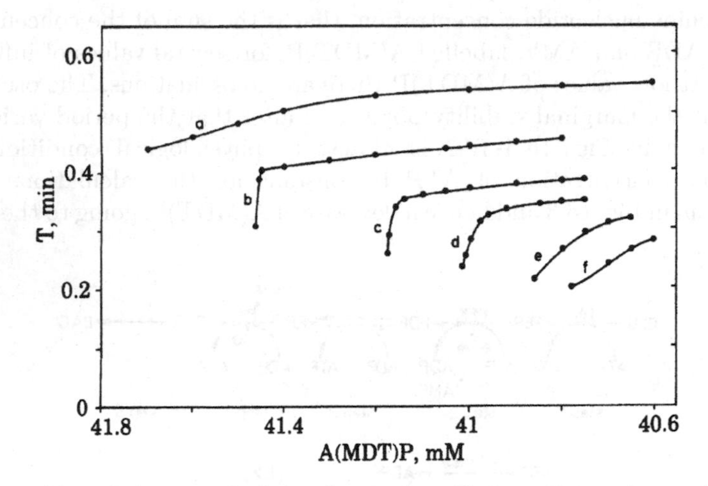

Fig. 16.5. Plot of the period of the chemical oscillation, T, vs. the total adenine nucleotide concentration for various values of influx conditions. The points of marginal stability, the transitions from non-oscillatory to oscillatory conditions, are on the extreme *left* of each curve. From [7]

Fig. 16.6, implies an increase in the ATP/ADP ratio. This is shown explicitly in Fig. 16.6, taken from [7], where we plot the average ratio of ATP/ADP for the oscillatory system, averaged over one oscillation, divided by that ratio in the unstable stationary state within the oscillatory system, vs. A(MDT)P.

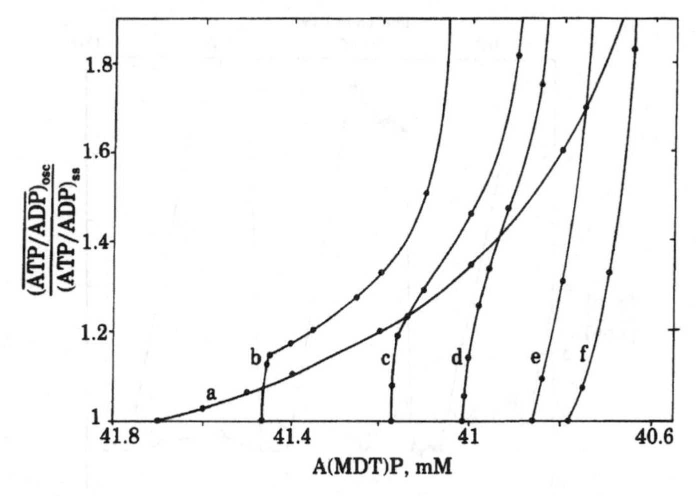

Fig. 16.6. Dependence of the average ratio of ATP/ADP in the oscillatory regime divided by that ratio in the stationary state. From [7]

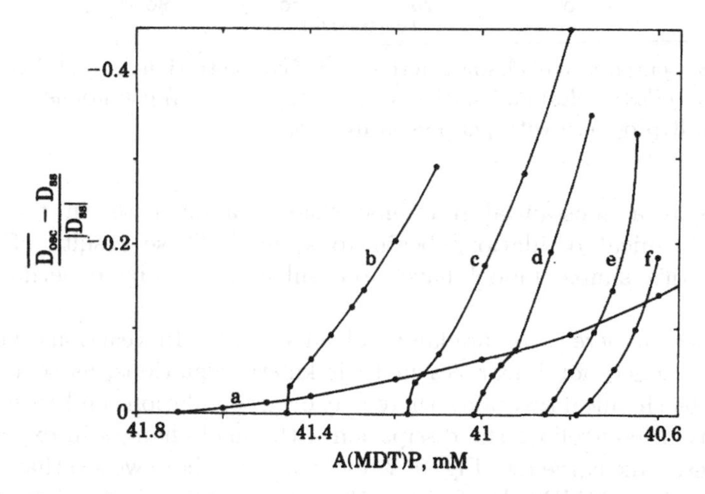

Fig. 16.7. Dependence of the average dissipation in the oscillatory regime on the overall reaction of ADP to ATP on the total adenine nucleotide concentration. From [7]

One of the curves in Fig. 16.6 is for given influx conditions. Hence when the system becomes oscillatory there is an increase in the average ATP to ADP ratio, that is an increase in the concentration of the ATP which is energy-rich compared to ADP. This in turn implies that on the average a larger fraction of the Gibbs free energy change for the overall conversion of ADP into ATP in the reaction mechanisms shown in Fig. 16.3 is channelled towards ATP and hence a smaller fraction towards dissipation. We show this important result in Fig. 16.7.

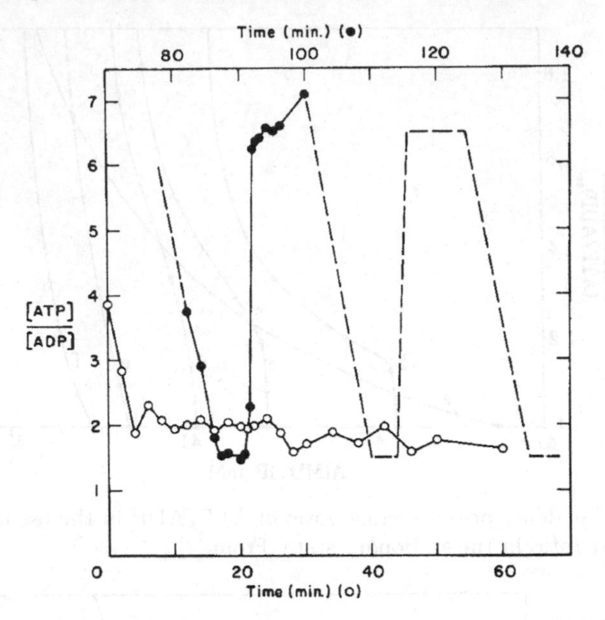

Fig. 16.8. Comparison of changes in the ratio of concentrations of ATP/ADP during oscillatory (*filled circles*) and stationary state (*open circles*) conditions in glycolysis in muscle. Reproduced with permission from [8]

There is a substantial fractional change, a decrease, in the dissipation as chemical oscillations begin to appear. These results of calculations on this simple model have been substantiated by experiments, see Fig. 16.8.

The lesson here is of fundamental importance. In reactions with feedback, and hence non-linearities in their kinetic equations, as so frequently found in biochemical reactions, there may occur oscillations and with that the possibility of controlling the dissipation with small changes in experimental conditions. Take curve c in Fig. 16.7, for example, where we see that a change (a decrease) in A(MDT)P of about 5% can lower the dissipation by nearly 20%. Conversely, if the system is oscillatory, say curve c at A(MDT)P = 41, and the need for heat, that is an increase in dissipation, becomes more urgent than the need for ATP, then a 5% change (an increase) in A(MDT)P can achieve that increase in dissipation by 20%.

In oscillatory reactions the Gibbs free energy of the overall reaction is oscillatory, the rate is oscillatory, and so is the dissipation. There now appears a new quantity, the phase relation between the oscillatory Gibbs free energy and the oscillatory rate. The dissipation, which is the product of these two quantities (see (13.1) and (13.6) in Chap. 13) depends critically on this phase relation and is controlled by it. There is an analogy here between DC and AC circuits on the one hand, and reactions in time-independent stationary states and in oscillatory states, on the other. We return to that interesting possibility of an 'alternating current chemistry' in Chap. 17 when we discuss experiments on reactions in which we determine these phase relations.

In [9], an analysis of the results of decreased dissipation with the onset of oscillations in the mechanism of glycolysis is presented, Fig. 16.3, as shown in Fig. 16.7. This reaction mechanism can be thought of as a combination of the phophofructose kinase (PFK) reaction and the pyruvate kinase (PK) reaction.The PFK subsystem by itself shows sustained oscillations in ranges of parameters of the order of experimental values. The (PK) reaction, however, is in a stable stationary state, which on perturbation away from that state relaxes back to the stationary state with an oscillatory component (a stable focus). If, for a given set of kinetic parameters, the period of oscillation of the (PFK) reaction is of the same order of magnitude as the oscillatory relaxation of the (PK) reaction, then the onset of oscillatory behaviour in the entire reaction mechanism past marginal stability forces a tuning of the (PFK) reaction period to that of the (PK) reaction. Thus the (PK) reaction tunes the frequency of the primary oscillophor, the (PFK) reaction and the species involved in this reaction, so that a resonance response results in the (PK) reaction. It is this resonance response that leads to decreased dissipation and an increase in the efficiency of the reaction to produce ATP.

In [10], we showed, as additional evidence, that substantial changes in dissipation may occur when the reaction mechanism in Fig. 16.3 is driven by a periodic input of glucose concentration. Within certain ranges of the frequency of the imposed periodic input of glucose, entrainment of the reaction mechanism may take place (see the discussion near Figs. 16.1 and 16.2) as shown in Fig. 16.9.

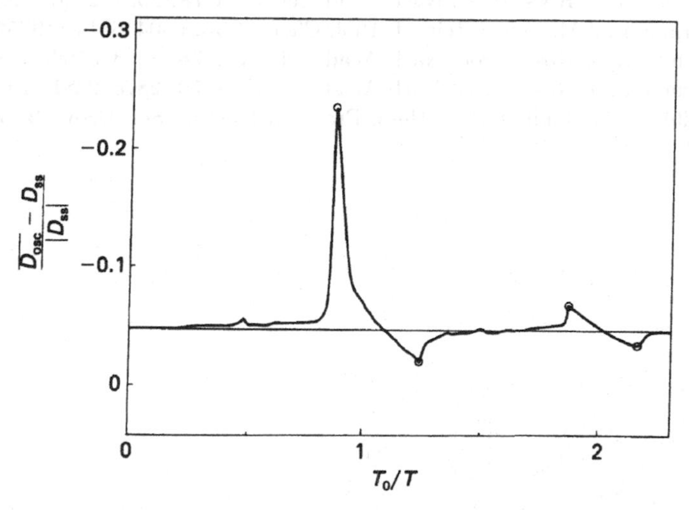

Fig. 16.9. Change in dissipation in entrainment bands. See text following the figure. From [10]

In Fig. 16.9 we plot the dissipation in an oscillatory state, averaged over one oscillation, minus the dissipation in the unstable stationary state from which the oscillations arise, divided by that same dissipation. T_0 and T are the periods of the self-sustained oscillation and the period of the external perturbation, respectively. The two circles on the curve near $T_0/T = 1$ indicate the extent of this fundamental entrainment band, and the two circles near $T_0/T = 2$ indicate the extent of this sub-harmonic entrainment band. The dissipation varies substantially in the fundamental entrainment band, and varies in the sub-harmonic entrainment band, but less so. Fundamental entrainment occurs in the range $0.85 < T_0/T < 1.25$, which agrees well with the experimental findings $0.83 < T_0/T < 1.42$, [11].

Acknowledgement. Based in part on [7,9].

References

1. E.E. Sel'kov, Eur. J. Biochem **4**, 79–86 (1968)
2. R.J. Field, M. Burger, *Oscillations and Traveling Waves in Chemical Systems* (Wiley, New York, 1985)
3. I. Epstein, J.A. Pojman, *An Introduction to Nonlinear Chemical Dynamics: Oscillations, Waves, Patterns, and Chaos* (Oxford University Press, New York, 1998)
4. P. Rehmus, J. Ross, *Oscillations and Traveling Waves in Chemical Systems* (Wiley, New York, 1985) Chap. 9 pp. 287–332
5. P.H. Richter, P. Rehmus, J. Ross, *Prog. Theor. Phys.* **66**, 385–405 (1981)
6. K. Tomita, T. Kai, F. Hikami, *Prog. Theor. Phys.* **57**, 1159–1177 (1977)
7. Y. Termonia, J. Ross, Proc. Natl. Acad. Sci. USA **78**, 2952–2956 (1981)
8. K. Tornheim, J.M. Lowenstein, J. Biol. Chem. **250**, 6304–6314 (1975)
9. Y. Termonia, J. Ross, Proc. Natl. Acad. Sci. USA **78**, 3563–3566 (1981)
10. Y. Termonia, J. Ross, Proc. Natl. Acad. Sci. USA **79**, 2878–2881 (1982)
11. A. Boiteux, A. Goldbeter, B. Hess, Proc. Natl. Acad. Sci. USA **72**, 3829–3833 (1975)

17

Three Applications of Chapter 16

In this chapter, we study the efficiency of three biochemical reactions: two systems with calculations, one with experiments. All three cases are applications of achieving variable efficiency and dissipation by means of externally forced oscillatory reactions, discussed in Chap. 16.

17.1 Thermodynamic Efficiency in Pumped Biochemical Reactions

We investigate the problem of establishing a concentration gradient, say across a membrane. This establishment requires energy, for example from ATP. In turn the gradient may be used to do work in the surroundings of the system. The subject is closely related to the issue of the efficiency of biological pumps, such as pumps for sodium ions, potassium ions, and protons.

Let us analyze a possible experimental system, shown in Fig. 17.1.

White light of a given intensity illuminates all of subvolume A and is absorbed by a chromatophore, Chr. This absorbed light provides the energy for the formation of ATP from ADP and phosphate, P_i. The ATP is used in the enzymatic PFK reaction to yield FDP. By means of this production, a gradient of FDP is established with the concentration of FDP in section A exceeding the concentration of FDP in the section B. Because of the decrease in F6P in A, a gradient is also established in F6P with $(F6P)_A < (F6P)_B$. Diffusion occurs across the membrane of both FDP and F6P, but if the rate of formation of FDP is sufficiently fast then a stationary state gradient in FDP and F6P will be formed. FDP reacts in B to form F6P.

The energy of the light absorbed by the chromatophore in A can establish and maintain a chemical potential difference in FDP and F6P, which may be used to do work. For discussions of the use of light to drive systems away from equilibrium see [2–6].

The deterministic kinetic equations for this system are highly nonlinear because of both the chromatograph reaction and the PFK reaction. (These

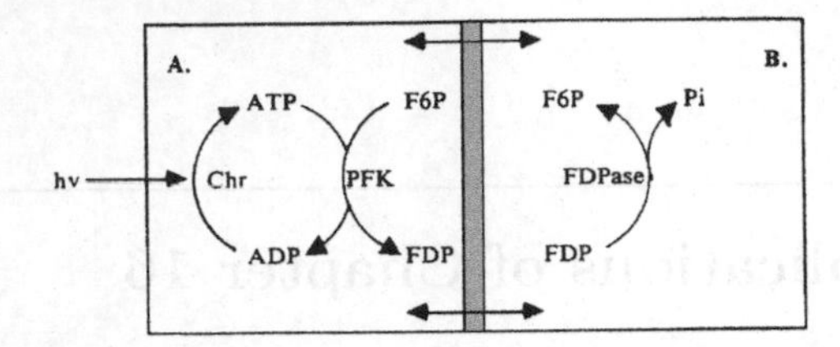

Fig. 17.1. A possible experiment for establishing a concentration gradient across a membrane. For description, see the text following the figure (from [1])

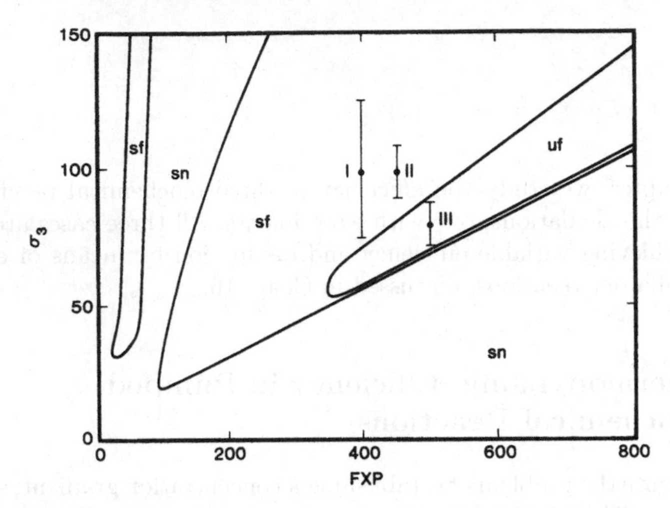

Fig. 17.2. Plot of a parameter σ_3, proportional to the light intensity, vs. FXP, the sum of the concentrations of F6P and FDP in compartments A and B (from [1])

equations are given in detail in [1].) We reproduce here only a plot of the dynamical domains of the stationary state solutions of the kinetic equations obtained by numerical analysis. These are shown in Fig. 17.2.

There are several different dynamic domains possible at stationary state: stable nodes, sn; stable foci, sf; and unstable foci, uf, that is stable oscillations.

We can study this system with a constant input of light intensity or an oscillatory input of light intensity with a given frequency and amplitude. Three cases, marked I, II, and III, are shown in Fig. 17.2 for which calculations were made of the efficiency of energy transduction from light to work: two of these are within a region of a stable focus, and one within a region of a limit cycle.

We need to define this efficiency of energy transduction and do so by defining first the power input for a given light intensity I_0

$$P_{\text{in}} = I_0 \tag{17.1}$$

There can be a choice here whether we take the power input to be given by the total light intensity or by the light intensity absorbed by the chromatophore (17.6). The power output of the system in Fig. 17.1 is taken to be the PFK reaction

$$P_{\text{out}} = \Delta G_1 J_{\text{PFK}},$$
$$\text{with } \Delta G_1 = \Delta G_{0,1} + RT \ln\left(\frac{\text{FDP}_\text{A}}{\text{F6P}_\text{A} \cdot \text{Pi}_\text{A}}\right) \tag{17.2}$$

in which J_{PFK} is the rate of the PFK reaction. The power output can produce work in unit time, the work obtained from transferring F6P to FDP.

The experimental system, Fig. 17.1, can be viewed either as an energy transduction or a power transduction engine in which light energy per unit time is converted to Gibbs free energy per unit time. We define the thermodynamic efficiency of this engine as the ratio of the power output to the power input

$$\eta = \frac{\langle P_{\text{out}} \rangle}{\langle P_{\text{in}} \rangle}, \tag{17.3}$$

where the angle brackets denote an average over one period oscillation of the light input. We wish to compare this efficiency under stationary state conditions in which the light input, the illumination, is constant to that under conditions in which the light intensity varies sinusoidally with a given amplitude and frequency. In Fig. 17.3 we show ratios of efficiency vs. ratios of frequency for three cases indicated in Fig. 17.2. For the two cases I and II, the system under steady illumination is in a stable focus, and for case II, the system is in stable limit cycle. For cases I and II (stable focus), the increases in the efficiency ratio occur at and near the frequency of relaxation of the stable focus and reach as high as 10%. For case III, a stable limit cycle, there are small increases in the efficieny near the 1:3, 1:2, and 2:1 entrainment bands, and a decrease near the 1:1, the fundamental entrainment band.

Changes in efficiency with oscillatory input may come about due to phase shifting of fluxes and forces, in this case the rate of the PFK reaction and the

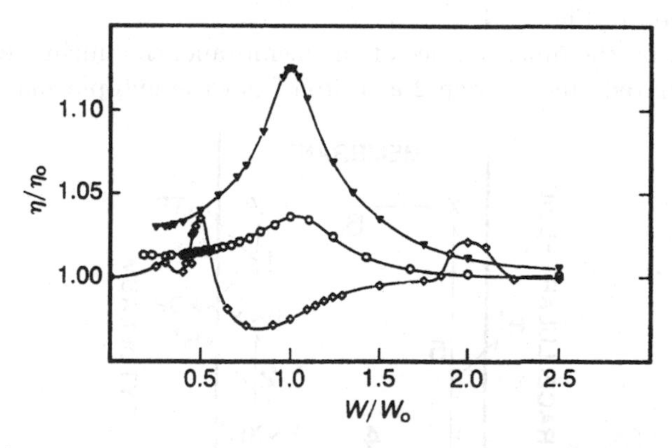

Fig. 17.3. Plot of the relative efficiency, the ratio of the efficiency of the system with oscillatory illumination to that with constant illumination, vs. the ratio of the frequency of illumination to the frequency of the autonomous system for the 3 cases, I, II, III, shown in Fig. 17.2. The amplitude of the oscillatory component of the illumination is 25% for the top curve and 10% for the lower two curves. From [1]

Gibbs free energy change of that reaction; due to changes in the magnitude of response upon oscillatory input; and due to changes in average values of the flux and the force. These points will be shown with experimental evidence in Sect. 17.3.

17.2 Thermodynamic Efficiency of a Proton Pump

Protons are pumped in living systems to establish a proton gradient, and the energy necessary for this pumping is frequently provided by the hydrolysis of ATP, in which ADP and phosphate are formed [7]. In this section, we study a model of a proton pump found in the plasma membrane of plants [8–12] and include the coupling of potassium and calcium ion transport. As in prior examples, we calculate the thermodynamic efficiency [13] of the proton pump with a constant influx of ATP and compare that to the thermodynamic efficiency with an oscillatory influx of ATP, the average of which is the same as the constant concentration of ATP.

In formulating the mechanism of the proton pump, we require the presence of certain nonlinearities to find the possibility of changing the dissipation, or the efficiency, with an oscillatory input of ATP. The minimum elements of a proton pump, although nonlinear, lead only to monotone relaxation kinetics, and thus only to decreases in efficiency upon imposition of an oscillatory influx of ATP. However, by including the coupling of other ion transport processes, such as those of potassium and calcium, the mechanism of the proton pump behaves like a damped oscillator, which has been observed in experiments (17.12). With that property a change in efficiency, increases and decreases, is feasible with an oscillatory influx of ATP.

The reaction sequence of a model of the cyclic, ATP driven proton pump [11,13] is given in Fig. 17.4.

In step 1, on the inner surface of the membrane, the uncharged enzyme X is phosphorylated, and in step 2 a proton from the cytoplasma binds to the

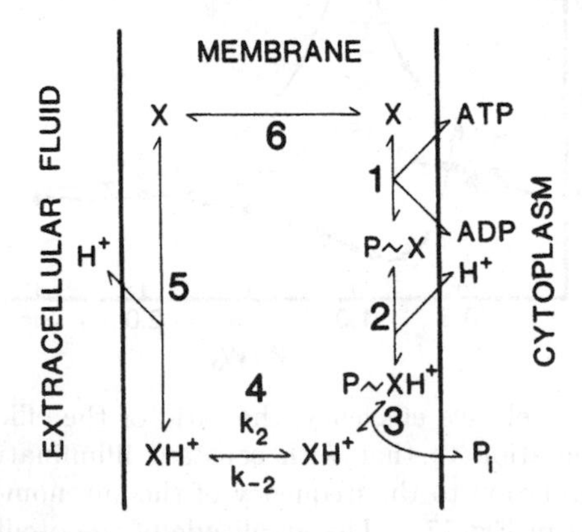

Fig. 17.4. Model reaction sequence for a proton pump. See the text following the figure (from [13])

phosphorylated enzyme. In step 3, the enzyme releases inorganic phosphate, retains the Gibbs free energy from the hydrolysis, and then undergoes conformational changes (or transit) in step 4. In step 5, a proton is transferred to the extracellular fluid and then returns in step 6 to the initial state for another cycle. The carrier concept is a convenience.

The rate coefficients k_2 and k_{-2} in step 4 are potential dependent and assumed to be given by

$$k_{\pm 2} = k_{\pm 20} \exp\left[-\left(\pm F\Delta\psi/2RT\right)\right] \tag{17.4}$$

where the symbols are defined after (17.5).

The deterministic kinetic equations for this system are given in [13], but we need not repoduce them here; they are nonlinear and for experimental values of the rate coefficients represent a damped oscillator.

The thermodynamic efficiency of the proton pump is defined by the equation

$$\eta = \langle \left(\mu_{\mathrm{H_o^+}} - \mu_{\mathrm{H_i^+}}\right) j_c + \Delta\psi I_{\mathrm{p}} \rangle / \langle -\Delta G_{\mathrm{hyd}} j_{\mathrm{hyd}} \rangle . \tag{17.5}$$

The angle brackets indicate a time average over one period of the oscillatory input of ATP. The (...) bracket gives the difference in the chemical potential of the protons across the membrane, which is equal to the work required to move protons against the proton concentration gradient. The rate at which protons are pumped into the extrcellular fluid is j_c; $\Delta\psi$. I_{p} is the work required to move charged particles against the membrane potential times the rate at which the proton pump performs this process; ΔG_{hyd} is the Gibbs free energy change for the hydrolysis of ATP; and j_{hyd} is the rate of hydrolysis.

The efficiency of the proton pump is defined as the total chemical and electrical work produced in unit time divided by the power made available from the hydrolysis of ATP, that is the Gibbs free energy change of hydrolysis in unit time

$$j_{\mathrm{c}} = j_{\mathrm{hyd}} = k_1 [\mathrm{ATP}] (X_{\mathrm{T}} - X_{\mathrm{in}}) - k_{-1} [\mathrm{ADP}] X_{\mathrm{in}} \tag{17.6}$$

with

$$-\Delta G_{\mathrm{hyd}} = -\Delta G^{\circ} + RT \ln([\mathrm{ATP}]/[\mathrm{ADP}]\,[\mathrm{P_i}]), \tag{17.7}$$

where X_{T} is the total ATPase concentration, X_{in} is the enzyme concentration within the membrane, P_{i} is the concentration of inorganic phosphate, and ΔG^0 is the standard Gibbs free energy change of the hydrolysis of ATP.

The efficiency for this model of a proton pump has been calculated [13] for different parameters in the stationary state mode and for a range of frequencies and amplitudes of ATP influx in the oscillatory mode. In Fig. 17.5, from [13], we plot the ratio of the efficiency in the oscillatory mode to that in the stationary mode vs. the ratio of the frequency of the ATP oscillation to the frequency of the autonomous system, the damped oscillator.

The relative efficiency is clearly a function of the frequency of the ATP oscillatory influx, with increases in certain ranges of frequency and decreases

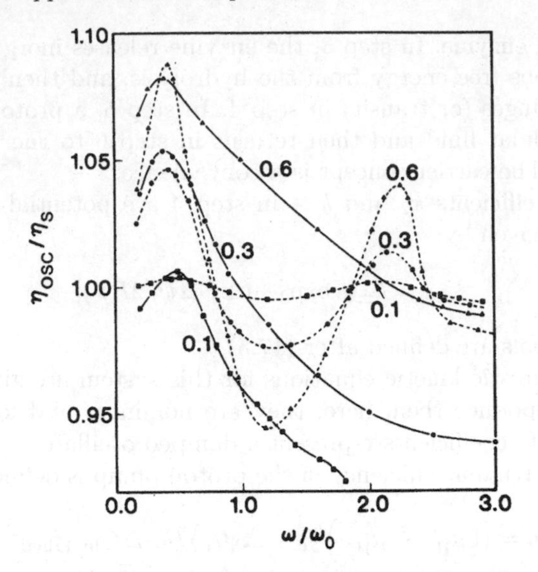

Fig. 17.5. The ratio of the efficiency in the oscillatory mode to that in the stationary state mode vs. the ratio of the frequency of the ATP oscillation to the relaxation frequency of the autonomous system. The amplitudes of perturbation are: 0.1 [ATP] *squares*; 0.3 [ATP] *circles*; 0.6 [ATP] *triangles*

in others. Changes in efficiency increase with the amplitude of perturbation. The solid curves are for one set of parameters, and the broken curves for another set (see [13]). For the first set parameters, solid curves with lower potassium and calcium conductances, there is a single range of frequencies with increased efficiencies with a maximum located at about $\omega/\omega_0 = 0.5$. For the second set of parameters, broken curves with higher potassium and calcium conductances, the system is near a Hopf bifurcation, a transition in the autonomous system from damped oscillations to undamped oscillations, which is a limit cycle. As this bifurcation is approached, a second maximum appears in the relative efficiency located at about $\omega/\omega_0 = 2$. For more details see [13].

Region at higher frequencies have not been investigated.

More on the subject of efficiencies are presented in the next section.

17.3 Experiments on Efficiency in the Forced Oscillatory Horse-Radish Peroxidase Reaction

In this section, we turn to experiments on the horse radish peroxidase (HRP) reaction [14, 15], which is the oxidation of nicotinamide adenine dinucleotide (NADH) catalyzed by HRP

$$\mathrm{NADH} + \mathrm{H}^+ + \frac{1}{2}\mathrm{O}_2 \xrightarrow{\mathrm{HRP}} \mathrm{NAD}^+ + \mathrm{H}_2\mathrm{O} \tag{17.8}$$

NAD^+ is recycled to NADH by glucose-6-phosphate dehydrogenase (G6PDH)

$$\begin{aligned} NAD^+ + G6P &\xrightarrow{G6PDH} 6PGL \\ &+ NADH + H^+ . \end{aligned} \tag{17.9}$$

This reaction under fixed pressure of oxygen is oscillatory, a limit cycle.

First, we show experimentally that the mode of supply of reactants, steady or oscillatory, with the average concentration of the oscillatory mode equal to that of the steady mode, can alter the stationary state concentrations of the reaction. Second, we show that a change in the stationary state concentration of a nonequilibrium chemical reaction is generally equivalent to a change in the average rate of the reaction. If an oscillatory input flux of a reactant is applied then there will be a temporal variation of the Gibbs free energy and the rate. Third, we show that there may be a phase shift of the temporal variation of ΔG with that of the temporal variation of the rate of the reaction, which may result in changes in dissipation, power output, and efficiency of the process.

The apparatus used for these experiments, shown in Fig. 17.6, consists of a reaction vessel, means of measuring the NADH (absorption) and O_2 (with an oxygen microelectrode) concentrations, and devices for a constant and oscillatory influx of O_2.

The Gibbs free energy change of the reaction is calculated from the measurements with the equation

$$\Delta G = \Delta G^o + RT \ln \frac{NAD^+}{NADH\, [O_2]^{1/2}} . \tag{17.10}$$

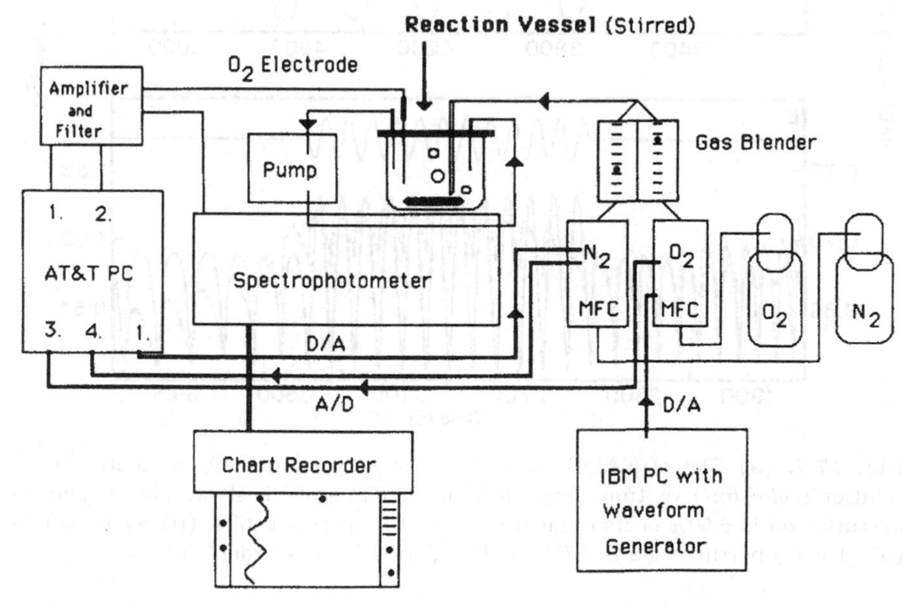

Fig. 17.6. For description see text (from [14])

The rate of the HRP reaction, J_{HRP}, is determined from the slope of the measurement of the NADH concentration as a function of time, since we have

$$\frac{d\mathrm{NADH}}{dt} = J_{\mathrm{G6PDH}} - J_{\mathrm{HRP}}, \tag{17.11}$$

where J_{G6PDH} is the known rate of regeneration of NADH from NAD^+. The dissipation of the HRP reaction is calculated from the equation

$$D = \Delta G \left(J_{\mathrm{HRP}}\right). \tag{17.12}$$

Some of the experimental results obtained in this study [14, 15] are shown in the next four figures. In Fig. 17.7 we present a plot of the NADH absorption (concentration), dashed line, and O_2 concentration in solution, solid line, in mM/l (scale on the ordinate at the right). The oxygen input, stationary or oscillatory, is shown in the top line. In Fig. 17.7 we see that changes in the NADH absorption and O_2 concentration in solution occur

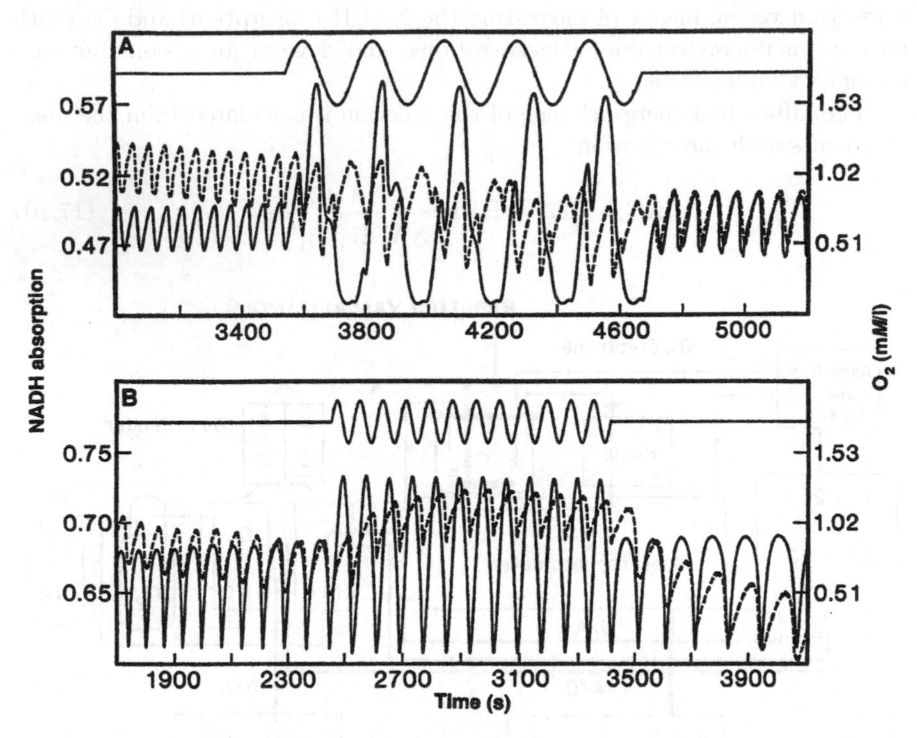

Fig. 17.7. **(a)** Plot of NADH (*dashed line*) absorption and O_2 concentration in solution (*solid line*) vs. time. Oxygen input, not to scale, is shown above. The O_2 perturbation is ±50% of its original value and its period is 60 s. **(b)** Same plot as **(a)**. The O_2 perturbation is ±75% and its period is 230 s (from [15])

when the O_2 gas influx changes from stationary to oscillatory, and hence the average concentrations of these two species differ in the region of oscillatory influx of O_2 from that in the stationary influx.

We use (17.10, 17.11) to calculate from the data in Fig. 17.7 the Gibbs free energy change and the rate of the HRP reaction, and these are shown in Fig. 17.8. Again there are changes in the oscillatory influx compared with the stationary influx, changes in average values of these quantities, and a change in their relative phase.

The change in relative phase of the rate and the Gibbs free energy change is given in more detail and clearly shown in Fig. 17.9.

These changes imply a change in the dissipation of the reaction, see (17.12), and hence in the efficiency of the reaction as shown in Fig. 17.10.

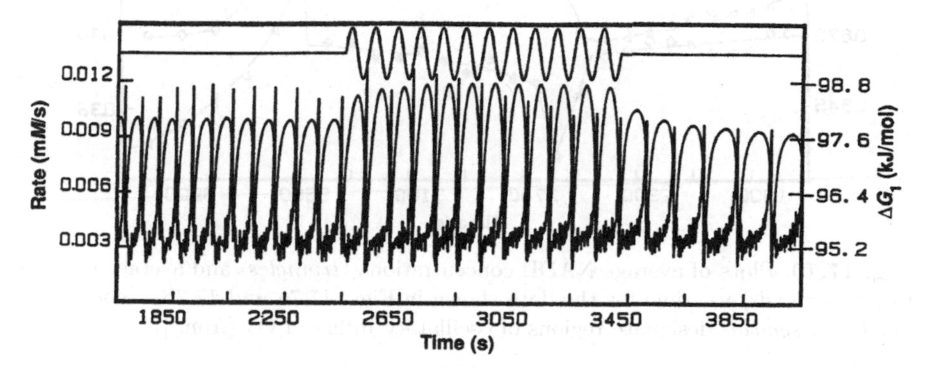

Fig. 17.8. Plot of ΔG_1 and the rate of the HRP reaction, (17.8), vs. time for the data plotted in Fig. 17.7a (from [15])

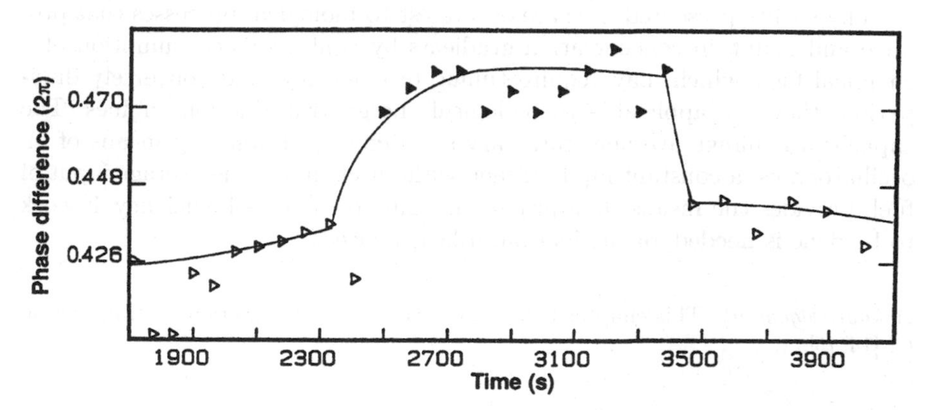

Fig. 17.9. Plot of the phase difference between the Gibbs free energy change and the rate of the HRP reaction in units of 2π for the data presented in Fig. 17.7a. The *darkened symbols* denote the region of oscillatory influx of O_2 (from [15])

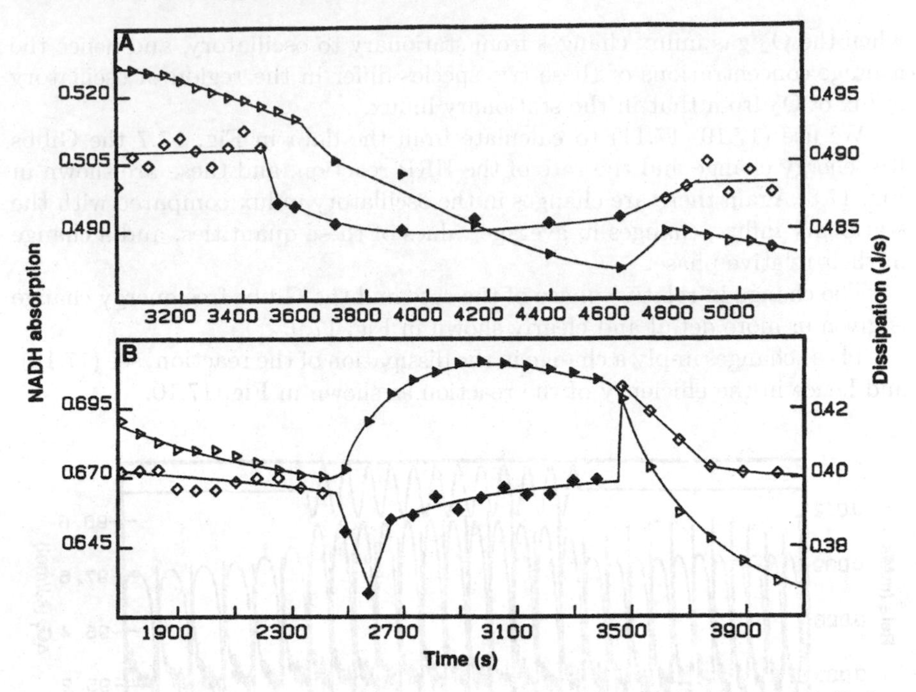

Fig. 17.10. Plots of average NADH concentrations (*triangles*) and average dissipation (*diamonds*) vs. time for the data shown in Figs. 17.7a and 17.7b, respectively. *Darkened symbols* designate regions of oscillatory influx of O_2 (from [15])

These experimental results substantiate the claims made at the beginning of this section (the paragraph following (17.9)).

The results presented here are of interest to biological processes that produce and maintain concentration gradients by continuous consumption of a chemical fuel, which may require tuning of efficiency, and conversely dissipation; they are applicable to biological energy transduction engines. The capacity to change average stationary state concentrations, by means of an oscillatory, vs. a constant input of fuel, without changing the average input of fuel, provides the means to adjust to changing demands of efficiency, if work to be done is needed, or dissipation, if heat is needed.

Acknowledgement. This chapter is based on: Section A. [1]; section B. [13]; section C. [14,15].

References

1. J.F. Hervagault, J.G. Lazar, J. Ross, Proc. Natl. Acad. Sci. USA **86**, 9258–9262 (1989)
2. J.G. Lazar, Studies on the thermodynamic efficiency and kinetics of a non-linear biochemical reaction subject to an external periodic perturbation. Ph.D. Thesis 1989, Stanford University, Stanford, CA
3. C.L. Creel, J. Ross, J. Chem. Phys. **65**, 3779–3789 (1976)
4. E.C. Zimmerman, J. Ross, J. Chem. Phys. **80**, 720–729 (1984)
5. E.C. Zimmerman, J. Ross, J. Chem. Phys. **81**, 1327–1336 (1984)
6. Y. Termonia, J. Ross, J. Chem. Phys. **74**, 2339–2345 (1981)
7. L. Stryer, *Biochemistry*, 4th ed. (W.H. Freeman and Company, New York, 1995)
8. J.P. Dufour, A. Goffeau, Eur. J. Biochem. **105**, 145–154 (1980)
9. R. Serrano, FEBS Letters **156**, 11–14 (1983)
10. J. Slavik, A. Kotyk, Biochim. Biophys. Acta **766**, 679–684 (1984)
11. C.L. Slayman, W.S. Long, D. Gradmann, Biochim. Biophys. Acta **426**, 732–744 (1975)
12. C.L. Slayman, 'Charge-transport characteristics of a plasma membrane proton pump' in *Membranes and Transport*, vol. 1 (Plenum Press, New York, 1982) pp. 485–490
13. M. Schell, K. Kundu, J. Ross, Proc. Natl. Acad. Sci. USA **84**, 424–428 (1987)
14. J.G. Lazar, J. Ross, J. Chem. Phys. **92**, 3579–3589 (1990)
15. J.G. Lazar, J. Ross, Science **247**, 189–192 (1990)

Part III

Stochastic Theory and Fluctuations in Systems Far from Equilibrium, Including Disordered Systems

18

Fluctuation–Dissipation Relations

We introduced the concepts of fluctuations and dissipation in Chap. 2, where we discussed the approach of a chemical system to a nonequilibrium stationary state; we recommend a review of that chapter. We restricted there the analysis to linear and nonlinear one-variable chemical systems and shall do so again in this chapter, except for a brief referral to extensions to multivariable systems at the end of the chapter. In Chap. 2 we gave some connections between deterministic kinetics, with attending dissipation, and fluctuations, see for example (2.33), which equates the probability of a fluctuation in the concentration X to the deterministic kinetics, see (2.8, 2.9). Here we enlarge on the relations between dissipative, deterministic kinetics, and fluctuations for the purpose of an introduction to the interesting topic of fluctuation dissipation relations. This subject has a long history, more than 100 years [1,2]; Reference [1] is a classical review with many references to fundamental earlier work. A brief reminder of one of the early examples, that of Brownian motion, may be helpful.

The dynamics of a Brownian particle of mass m velocity v is described by a Langevin equation [3]

$$m\mathrm{d}v/\mathrm{d}t + \gamma v = F(t), \tag{18.1}$$

where the first term is the force due to acceleration of the particle, the second term the frictional force on the particle, and $F(t)$ is a delta correlated Gaussian random force for which the first two moments are

$$\langle F(t)\rangle = 0, \quad \langle F(t)F(t')\rangle - 2\lambda\delta(t-t'), \tag{18.2}$$

in which λ is a parameter that gives the strength of the fluctuations that affect the Brownian particle. In a derivation due to Langevin, given in [3], it is shown that the friction coefficient γ in (18.1), which characterizes the dissipative process, and λ, which characterizes the fluctuations, are related by

$$\lambda\gamma = kT. \tag{18.3}$$

Equation (18.3) is a fluctuation–dissipation relation.

Now consider an one-variable open chemical system, either linear

$$A \underset{k_2}{\overset{k_1}{\leftrightarrow}} X \underset{k_4}{\overset{k_3}{\leftrightarrow}} B. \tag{18.4}$$

or nonlinear

$$A + 2X \underset{k_2}{\overset{k_1}{\leftrightarrow}} 3X, \quad X \underset{k_4}{\overset{k_3}{\leftrightarrow}} B. \tag{18.5}$$

We have defined a thermodynamic state function for the linear system in (2.6, 2.5), and for the nonlinear system in (2.13, 2.12), in each case in terms of species-specific affinities and the kinetic rates in the forward and reverse reactions. (We labeled these state functions there with ϕ and ϕ^*, respectively, but use here the common label ϕ). With two definitions, one for the net rate of the reaction

$$t(x) = t^+(x) - t^-(x) \tag{18.6}$$

and the other for the total rates in the forward and reverse reaction

$$D(x) = [t^+(x) + t^-(x)]/2 \tag{18.7}$$

we write a fluctuation–dissipation relation (not out of thin air)

$$t(x) = 2D(x)\tan\mathrm{h}\left[-\frac{1}{2Vkt}\left(\frac{\partial\phi}{\partial x}\right)\right] \tag{18.8}$$

$$t(x) = 2D(x)\tanh[-A(x)/2kT] \tag{18.9}$$

with the species specific affinity $A(x)$

$$A(x) = \mu_X(x) - \mu_X(x_s), \tag{18.10}$$

for the linear case, and

$$A(x) = \mu_X(x) - \mu_X(x^*) \tag{18.11}$$

for the nonlinear case. Equation (18.8) is derived in [4], and we can make it readily reasonable: substitute (18.6, 18.7) into (18.8), write out the $\tan\mathrm{h}$ term, $\tan\mathrm{h}\, x = [\exp x - \exp -x]/[\exp x + \exp -x]$, and you obtain consistently the derivative of (2.6) for the linear reaction mechanism, and (2.13) for the nonlinear reaction mechanism.

What are the advantages of the formulation, (18.8)? The term $t(x)$ is the net flux of the deterministic kinetics, (18.6), and the derivative of the state function ϕ is the species specific affinity, $(\mu_x - \mu^s{}_x)$, for the linear case, or $(\mu_x - \mu^*{}_x)$ for the nonlinear case, the driving force for the reaction toward a stationary state. Thus we have a flux-driving force relation. Second, the formulation is symmetric with respect to $t^+(x)$ and $t^-(x)$, which is not the case with other formulations. Third, the state function ϕ determines the probability distribution of fluctuations in x from its value at the stationary state, see (2.34). Further, as we shall show shortly, the term $D(x)$ is a measure of the strength of the fluctuations of the total number of reaction events (total in the forward and reverse reaction) and for both reasons (18.8) is a fluctuation–dissipation relation. We emphazise that (18.8) is nonlinear and holds for arbitrarily large fluctuations.

Another advantage of (18.8) is due to obtaining expressions easily for four regimes for different regions of the affinity and the dissipation.

1. *The fluctuation limited regime*
   For large positive and negative values of the affinity for which $\tanh \mathrm{y} \approx \pm 1$, (18.8) becomes
   $$t(x) \approx -2D(x) sign\ A(x) \tag{18.12}$$
   with
   $$sign\ A(x) = |A(x)|/A(x) \tag{18.13}$$
   The system is far from the stationary state and the magnitude of $A(x)$ plays no role. The sign of $A(x)$ determines the sense of the reaction, forward or backward. This case is irreversible in the sense of kinetics.
2. *The intermediate regime*
   Here the values of $A(x)$, in the range $2.9 > |A(x)|/2kT > 0.2$, and the value of $D(x)$ are both important and (18.8) cannot be further simplified.
3. *The dissipation limited regime*
   Here the species-specific affinity is small and $\tanh \mathrm{y} \approx \mathrm{y}$ and (18.8) reduces to
   $$t(x) = -D(x)A(x)/kT \tag{18.14}$$
   Although the net rate is proportional to $A(x)$, (18.14) is nonlinear because of the x dependence of D.
4. *The linear thermodynamic regime*
   If we linearize the net rate, the affinity, and the strength of the fluctuations $D(x)$ away from a stationary state then (18.8) reduces to
   $$t(x) = -D(x_s)A(x)/kT \tag{18.15}$$
   where the net rate and the affinity are linear functions of x.

We need to return to a promised interpretation of the quantity $D(x)$ in the fluctuation–dissipation relation, (18.8). Consider the concept of 'reaction event' [5], that is a generation or consumption of X in reactions (18.4) or (18.5). Let the total number of reaction events be labelled $q(t)$, which occur in the time interval $(0, t)$. Further we take $q(t)$ to be a nondecreasing *random* function of time even at a stationary state and at equilibrium. The variable q is a discreet time scale; the chemical reaction is a clock, the time being measured by the number of reaction events. Since the reaction events are independent the distribution of reaction events obeys Poisson statistics [6]. The first and second moments of the Poisson distribution are

$$\begin{aligned} \langle q(x,t) \rangle &= 2VD(x)t, \\ \langle \Delta q^2(x,t) \rangle &= 2VD(x)t. \end{aligned} \tag{18.16}$$

The second of (18.16) is similar to the Einstein equation for the mean square displacement of a Brownian particle in one dimension

$$\langle \Delta X^2 \rangle = 2Dt. \tag{18.17}$$

Here X is the displacement, in one variable, and D is the diffusion coefficient of the Brownian particles. Hence we can identify $D(x)$ in (18.16) as a probability diffusion coefficient in concentration space. The dispersion of q, the second equation in (18.16), is proportional to $D(x)$ and hence $D(x)$ is a measure of the strength of the fluctuations of q at a stationary state.

Further insight into the quantity $D(x)$ can be obtained by introducing the age τ of a fluctuation state, that is the time interval between the last transition $X \pm 1 \to X$ and the moment of observation. The age τ is determined by a succession of random events and hence is a random variable, and obeys a stochastic master equation ([4], p. 7273). From the stationary form of that equation we derive the relation

$$D(x) = 1/(2V\langle \tau(x) \rangle). \tag{18.18}$$

We see that the diffusion coefficient in concentration space decreases as the average of the age of a fluctuation state increases, and thus $D(x)$ can be viewed as a measure of the instability of the fluctuation state. (There is no connection here with the concept of macroscopic stability of a stationary state.)

The quantity $<\tau(x)>$ depends on the size of the system and for a macroscopic system it is a very small quantity that may not be accessible to direct measurement. However, we can make a connection to a macroscopic time scale that of the mean lifetime of an intermediate, x, for a stationary state, which we label $<\theta(x)>$, which is an observable. For a macroscopic stationary state, the mean lifetime of a given concentration of the intermediate is given by the ratio of the concentration x present in the system and its rate of disappearence

$$<\theta(x)> = x_{\mathrm{s}}/t^{-}(x_{\mathrm{s}}) \tag{18.19}$$

Since $t^{-} = t^{+}$ we have from the definition of $D(x)$, (18.7),

$$\langle \theta(x_{\mathrm{s}}) \rangle = x_{\mathrm{s}}/D(x_{\mathrm{s}}). \tag{18.20}$$

From (18.18, 18.20) we obtain

$$\langle \theta(x_{\mathrm{s}}) \rangle / \langle \tau(x_{\mathrm{s}}) \rangle = 2X_{\mathrm{s}}. \tag{18.21}$$

Thus the ratio of the chemical time scale, given by the mean lifetime, to the fluctuation time scale is proportional to the number of intermediates, $X_{\mathrm{s}} = Vx_{\mathrm{s}}$. For an appreciation of these time scales we make estimates for a slow and a fast reaction. We consider a macroscopic system with volume of $10^{-6}\,\mathrm{m}^3$. In the first case, we take the concentration to be $x_{\mathrm{s}} = 2.7 \times 10^{22}$ molecule m^{-3}, which at 0 °C corresponds to a partial pressure of X, ${p^{\mathrm{st}}}_{\mathrm{s}} = 10^{-3}$ atm. In the second case, we take the concentration to be $x_s = 1.3 \times 10^{22}$ molecule m^{-3} at the same partial pressure and at 300 °C. For the quantities discussed we obtain the estimates

$$\begin{aligned}
&\text{(a) For the slow process:}\\
&\langle\theta\rangle = 10^3\,\text{s}, \quad \langle\tau\rangle = 1.85 \times 10^{-14}\,\text{s},\\
&D = 2.7 \times 10^{19}\,\text{s}^{-1}\,\text{m}^{-3}.\\
&\text{(b) For the fast process:}\\
&\langle\theta\rangle = 10^{-2}\,\text{s}, \quad \langle\tau\rangle = 3.9 \times 10^{-19}\,\text{s}.\\
&D = 1.3 \times 10^{24}\,\text{s}^{-1}\,\text{m}^{-3}.
\end{aligned}$$

These estimates confirm our expectations that the mean age of a fluctuation is small indeed.

The analysis of fluctuation–dissipation relations goes beyond what we intend to present here. In a few sentences we indicate directions and give some references. The stochastic transition probability can be formulated in terms of a 'chemical' Lagrangian for which an explicit expression can be given for one-variable systems in the thermodynamic limit of large systems (many particles) [7]. The fluctuation–dissipation relation discussed earlier can be obtained with his formalism, and there is an important connection between the chemical Lagrangian and the excess work ϕ that determines the stochastic probability distribution, (2.34).

A thermodynamic approach to nonequilibrium fluctuations is given in [8]. A connection is made to the work of Greene and Callen [9]. A detailed comparison is presented with the work of Keizer, which is limited to Gaussian fluctuations around a stable stationary state, and therefore not well suited to predictions of relative stability of multiple stationary states.

The problem of fluctuation–dissipation relations in multivariable systems is analyzed in [15]; the mathematics needed for that task goes beyond the level chosen for this book, and hence only a brief verbal summary is presented. A statistical ensemble is chosen, which consists of a large number of replicas of the system, such as for example the Selkov model, each characterized by different composition vectors. There exists a master equation for this probability distribution of this ensemble, which serves as a basis for this approach; an analytical solution of this master equation is given in [15].

In statistical mechanics, the condition of microscopic time reversal (microscopic reversibility) [16, 18] expresses the invariance of the microscopic equations of evolution with respect to changing the sign of the time variable. For systems without solenoidal fields, the application of time reversal leads to the condition of detailed balance, which states that for each direct process there is a reverse process, and at equilibrium the rate of each direct process equals the rate of the reverse process.

A less restrictive condition than microscopic reversibility is that of mesoscopic reversibility [15], which follows from the assumption that the ensemble master equation is invariant with respect to a change of sign of the time variable. The condition of mesoscopic reversibility is introduced not for equilibrium but for stationary states far from equilibrium. For example for the Schlögl model

$$\begin{aligned} A_1 + 2X_1 &\rightleftharpoons 3X_1 \\ A_2 &\rightleftharpoons X_1 \end{aligned} \tag{18.22}$$

microscopic reversibility (detailed balance) requires

$$r_1^+ = r_1^-, r_2^+ = r_2^- \tag{18.23}$$

where the r's are rates of reaction in the forward (+) and the reverse (−) direction of the first and second step of (18.22), as indicated by subscripts. Mesoscopic balance requires the less restrictive condition

$$r_1^+ + r_2^+ = r_1^- + r_2^- . \tag{18.24}$$

Most chemical systems of interest do not obey either the condition of detailed balance or that of mesoscopic balance. Nevertheless, the condition of mesoscopic balance provides a useful reference state. Mesoscopic balance can be described by an extremum principle: if the contributions of different reactions to the total number of reaction events are constant then the dispersions of the net numbers of the reaction events have minimum values for mesoscopic reversibility. We can obtain relations that provide a measure of the extent of deviation from mesoscopic reversibility, which are proportional to the average values of the net numbers of reaction events. Within that framework, explicit expressions can be derived for fluctuation–dissipation relations for multivariable systems.

For a review of developments on this and related subjects see [19].

Acknowledgment. This chapter is base in part on [4,7,15].

References

1. M. Lax, Rev. Mod. Phys. **32**, 25–64 (1960); M. Lax, Rev. Mod. Phys. **38**, 359–379 (1966); M. Lax, Rev. Mod. Phys. **38**, 541–566 (1966)
2. F. Reif, Fundamentals of Statistical and Thermal Physics (McGraw-Hill, New York, 1965) Chapters 1, 15; S. Chandrasekhar, *Rev. Mod. Phys.* **15**, 1–89 (1943)
3. R.S. Berry, S.A. Rice, J. Ross, *Physical Chemistry*, 2nd edn. (Oxford University Press, New York, 2000), pp. 861–862
4. M.O. Vlad, J. Ross, J. Chem. Phys. **100**, 7268–7278 (1994)
5. M. Solc, Z. Phys. Chem. New Folge **92**, 1–10 (1974); M. Solc, Collect. Czech. Chem. Commun. **39**, 197–205 (1974); M. Solc, Collect. Czech. Chem. Commun. **46**, 1217–1222 (1981)

6. R.S. Berry, S.A. Rice, J. Ross, *Physical Chemistry*, 2nd edn. (Oxford University Press, New York, 2000) p 469
7. M.O. Vlad, J. Ross, J. Chem. Phys. **100**, 7279–7294 (1994)
8. M.O. Vlad, J. Ross, J. Chem. Phys. **100**, 7295–7309 (1994)
9. R.F. Greene, H.B. Callen, Phys. Rev. **83**, 1231–1235 (1951)
10. J. Keizer, *Statistical Thermodynamics of Nonequilibrium Processes* (Springer, Berlin, Heidelberg New York, 1987)
11. J. Keizer, J. Chem. Phys. **63**, 5037–5043 (1975); J. Keizer, J. Chem. Phys. **64**, 1679–1687 (1976); J. Keizer, J. Chem. Phys. **64**, 4466–4474 (1976); J. Keizer, J. Chem. Phys. **69**, 2609–2620 (1978)
12. J. Keizer, Phys. Rev. A **30**, 1115–1117 (1984)
13. J. Keizer, J. Chem. Phys. **82**, 2751–2771 (1985)
14. J. Keizer, O.K. Chang, J. Chem. Phys. **87**, 4064–4073 (1987); J. Keizer, J. Chem. Phys. **87**, 4074–4087 (1987); J. Keizer, J. Chem. Phys. **93**, 6939–6943 (1989)
15. M.O. Vlad, J. Ross, J. Chem. Phys. A. **104**, 3159–3176 (2000)
16. L. Onsager, S. Machlup, Phys. Rev. **91**, 1505–1512 (1953); S. Machlup, L. Onsager, Phys. Rev. **91**, 1512–1515 (1953)
17. N. Hashitsume, Prog. Theor. Phys. **8**, 461–476 (1952); N. Hashitsume, Prog. Theor, Phys. **15**, 369–413 (1956)
18. N.G. Van Kampen, *Stochastic Processes in Physics and Chemistry*, 2nd edn. (North-Holland, Amsterdam, 1992)
19. G.N. Bochkov, Yu. E. Kuzovlev, Physica **106A**, 443–479 (1981)

19

Fluctuations in Limit Cycle Oscillators

There are many examples of oscillatory systems: model chemical reactions, chemical systems, biochemical systems, etc. [1]. A well-known model is that of Selkov [2]

$$R \underset{k_2}{\overset{k_1}{\rightleftarrows}} X, \quad X + 2Y \underset{k_4}{\overset{k_3}{\rightleftarrows}} 3Y, \quad Y \underset{k_6}{\overset{k_5}{\rightleftarrows}} P, \tag{19.1}$$

where R (P) denotes reactant (product) with fixed concentrations, X and Y are two intermediate oscillating chemical species and the ks are rate coefficients in the deterministic rate equations. Oscillations of X and Y occur in given ranges of the kinetic rate coefficients and constraints of R and P. In a stochastic analysis based on a birth–death master equation there are fluctuations around the deterministic oscillatory trajectory. We investigate these fluctuations without entering into too much mathematics.

Quite generally, we write the deterministic dynamic equation for the chemical species x_i in the form

$$\frac{\mathrm{d}x_i}{\mathrm{d}t} = A_1 (x_1, \ldots, x_N) \ (N \geq 2) \tag{19.2}$$

and suppose that a solution exists which is a stable limit cycle. It is convenient to choose another set of variables

$$(\xi_1, \ldots, \xi_N)$$

and the transformation from the set x to the set ξ is given by

$$B_{ij} = \frac{\partial x_i}{\partial \xi_j}. \tag{19.3}$$

We want the co-ordinates ξ_i to be perpendicular to each other and achieve that with the orthogonal unitary transformation matrix

$$\sum_k B_{ik} B_{jk} = \delta_{ij}. \tag{19.4}$$

The dynamical equations in terms of the ξ_i variables are

$$\frac{\mathrm{d}\xi_i}{\mathrm{d}t} = \sum_j \left(B^{-1}\right)_{ij} A_j = \sum_j B_{ji} A_j . \tag{19.5}$$

Close to the deterministic limit cycle we choose one coordinate, say ξ_1, as the length along the limit cycle.

For a two-variable system the deterministic kinetic equations are

$$\begin{aligned} \frac{\mathrm{d}x_1}{\mathrm{d}t} &= A_1 \left(x_1, x_2\right), \\ \frac{\mathrm{d}x_2}{\mathrm{d}t} &= A_2 \left(x_1, x_2\right). \end{aligned} \tag{19.6}$$

We write x_1 and x_2 for the concentrations of the species X and Y, and then A_1 and A_2 in (19.6) for the Selkov model are

$$\begin{aligned} A_1 &= k_1 + k_4 x_2^3 - \left(k_2 + k_3 x_2^2\right) x_1, \\ A_2 &= k_6 + k_3 x_2^2 x_1 - \left(k_5 + k_4 x_2^2\right) x_2. \end{aligned} \tag{19.7}$$

The birth–death master equation in terms of the numbers of species X and Y is

$$\begin{aligned} \frac{\mathrm{d}P(X,Y)}{\mathrm{d}t} = {} & \Omega k_1 P(X-1,Y) + k_2 (X+1) P(X+1,Y) \\ & + \frac{k_3}{\Omega^2} (X+1)(Y-1)(Y-2) P(X+1,Y-1) \\ & + \frac{k_4}{\Omega^2} (Y+1) Y (Y-1) P(X-1,Y+1) \\ & + k_5 (Y+1) P(X,Y+1) + \Omega k_6 P(X,Y-1) \\ & - \left[\Omega k_1 + k_2 X + \frac{k_3}{\Omega^2} X Y^2 + \frac{k_4}{\Omega^2} Y^3 + k_5 Y + \Omega k_6\right] P(X,Y), \end{aligned} \tag{19.8}$$

In the stationary state the left-hand side of (19.8) is set to zero.

Numerical solution of the stationary state of (19.8) obtained by a Monte Carlo procedure is shown in Fig. 19.1, taken from [3].

The symbol Ω appears in the master equation, (19.8). The deterministic path of the stable limit cycle is located on the ridge of the crater. The exponential in the stationary distribution of the master equation in the eikonal limit is an excess work related to thermodynamic functions, see Chaps. 2–7.

In [3] the master equation is approximated by a Fokker–Planck equation, which is linearized close to the deterministic limit cycle trajectory; the probability distribution in the degree of freedom perpendicular the limit cycle trajectory becomes a Gaussian distribution. A comparison of the numerical (Monte Carlo) results with those of the Fokker–Planck equation is given in Fig. 19.2.

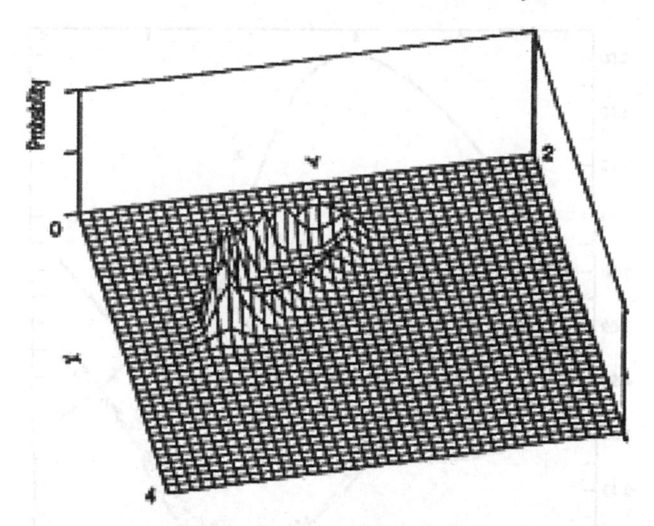

Fig. 19.1. Monte Carlo results for the stationary probability distribution for the Selkov model with the shape of a volcanic crater. The parameters are $k_1 = 1.0,\ k_2 = 0.2,\ k_3 = 1.0,\ k_4 = 0.1,\ k_5 = 1.105$ and $k_6 = 0.1$. The system has a stable deterministic limit cycle located on the ridge of the center. The symbol Ω denotes the effective dimensionless volume which scales the total number of molecules, taken to be $\Omega = 50,000$

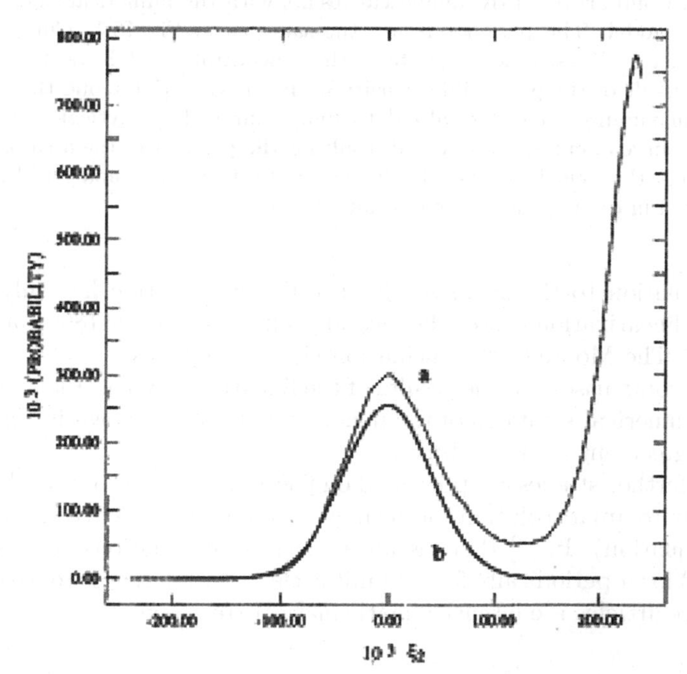

Fig. 19.2. Plot of the probability distribution in a cross section, tranverse to the ridge, for the Selkov model. (**a**) results of the Monte Carlo calculation (**b**) solution of the linearized Fokker–Planck equation. Taken from [3]

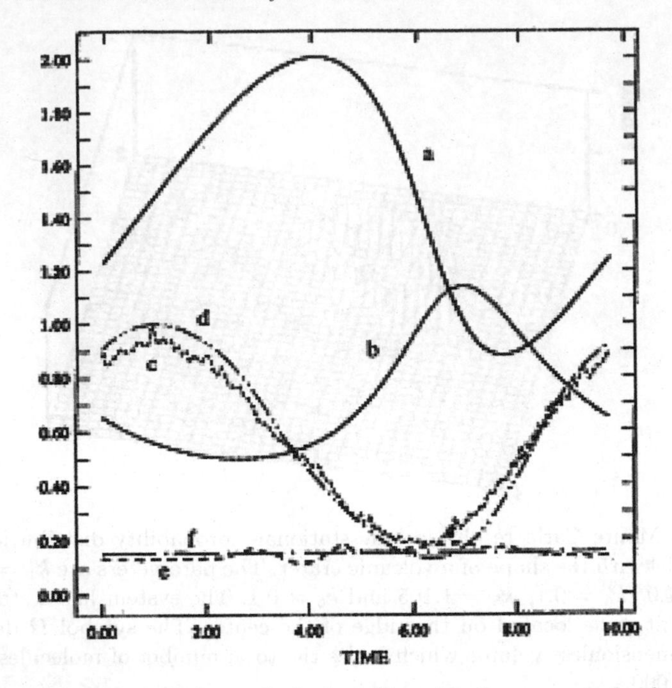

Fig. 19.3. Comparison of the analytical results with the numerical calculations on the Selkov model. The parameters are the same as in Fig. 19.1. Curve a: x, the concentration of X vs. time; curve b: y, the concentration of Y vs. time; curve c: numerical result of the probability density in the cross section along the limit cycle with the maximum value normalized to unity; curve d: analytical result for the same as in curve c; curve e: numerical result for the product of the area of the cross section times the velocity, which is almost constant; curve f: analytical result the same plotted in quantity as curve e. From [3]

The solution to the linearized Fokker–Planck equation has only a single peak; the linearization misses the second peak on the crater 180° opposite the first peak. The Monte Carlo calculation yields two peaks.

Other comparisons of the results of the linearized Fokker–Planck equation and the numerical solutions of the master equation are shown in Fig. 19.3.

The agreement is satisfactory.

In [4] further studies are presented on fluctuations near limit cycles, on the basis of approximate solutions of the master equation (rather than the Fokker–Planck equation). In [5] there is an analysis of fluctuations (the stochastic potential) for a periodically forced limit cycle, with references to earlier work. Both these articles are intensive mathematical treatments.

Acknowledgement. This chapter is based in part on [3].

References

1. I.R. Epstein, J.A. Pojman, *An Introduction to Nonlinear Chemical Dynamics: Oscillations, Waves, Patterns, and Chaos* (Oxford University Press, New York, 1998)
2. E.E. Sel'kov, Eur. J. Biochem **4**, 79–86 (1968)
3. M.I. Dykman, X.L. Chu, J. Ross, Phys. Rev. E **48**, 1646–1654 (1993)
4. W. Vance, J. Ross, J. Chem. Phys. **105**, 479–487 (1996)
5. W. Vance, J. Ross, J. Chem. Phys. **108**, 2088–2103 (1998)

20

Disordered Kinetic Systems

In the usual mass action chemical kinetics the rate coefficients are parameters with fixed values; these values may change with temperature, pressure, and possibly ionic strength for reactions among ions. In the field of disordered kinetics we broaden the study to systems in which the rate coefficients may vary. For some prior reviews on disordered kinetics, see [1–5].

Rate coefficients may vary due to environmental fluctuations and there are two categories of disorder: static and dynamic. In systems with static disorder the fluctuations of the environment are frozen and one fluctuation, once it occurs, lasts forever. For these systems, the fluctuations are introduced in the theoretical description by using random initial or random boundary conditions. (Thermal fluctuations are usually too small to be considered.) A typical example of a chemical reaction in a system with static disorder is a combination of an active intermediate in radiation chemistry in a disordered material, such as the sulphuric acid glass [6, 7]. The radiation of the active intermediate produces a reaction and the rate of that reaction differs at different sites in the glass. In systems with dynamical disorder the structure of the environment changes as the reaction progresses and the rate coefficients are random in time. An example of dynamic disorder is that of an enzyme in which a catalyzed reaction takes place at the active site of the enzyme and the rate of that reaction may depend on the configuration of the enzyme. As that configuration changes in time so does the rate coefficient of the catalytic reaction [8, 9].

The same system can display both types of disorder, depending on external conditions. For example, in the case of protein–ligand interactions [8, 9], the reaction rates are random because a protein can exist in many different molecular conformations, each conformation being characterized by a different reaction rate. At low temperatures, the transitions among the different conformations can be neglected, and the system displays static disorder. For higher temperatures, however, the transitions among the different conformations cannot be neglected, and the system displays dynamical disorder.

The chemical processes occurring in both types of disordered systems have the interesting property that fluctuations of the environment have a fundamental influence on the kinetic behaviour of the system and can lead to the substantial modification of the time dependence of the concentrations of the different chemicals. In contrast, in the case of chemical fluctuations described by a master equation the contribution of fluctuations for macroscopic systems is negligible. The qualitative difference between the fluctuations in these two types of systems can be investigated by studying the relative fluctuations of the number of molecules N_Ω of a chemical in the limit of very large volumes, $\Omega \to \infty$. The relative fluctuations $\rho_m(\Omega)$ of different orders on N_Ω are defined as ratios between the cumulants of N_Ω, $\langle\langle[N_\Omega]^m\rangle\rangle$, where $m = 2, 3, \ldots$ and the successive orders of the corresponding average value $\langle N_\Omega\rangle$

$$\rho_m(\Omega) = \langle\langle[N_\Omega]^m\rangle\rangle / \left(N_\Omega\right)^m . \tag{20.1}$$

In the eikonal approximation for ordered systems without environmental fluctuations all cumulants $\langle\langle[N_\Omega]^m\rangle\rangle$ are proportional to the volume Ω of the system in the limit $\Omega \to \infty$ [10],

$$\langle N_\Omega\rangle \sim \Omega, \langle\langle[N_\Omega]^m\rangle\rangle \sim \Omega, m = 2, 3, \ldots \text{ as } \Omega \to \infty, \tag{20.2}$$

and therefore all relative fluctuations tend to zero in the thermodynamic limit

$$\rho_m(\Omega) \sim \Omega^{-(m-1)}, m = 2, 3, \ldots \text{ as } \Omega \to \infty. \tag{20.3}$$

The fluctuations of this type are called non-intermittent; they are commonly encountered in statistical mechanics and have a negligible contribution to the behaviour of macroscopic systems. For these types of processes in the limit of large volumes, the average values of concentrations computed by taking fluctuations into account are practically identical to the values computed by neglecting the fluctuations and solving the deterministic kinetic equations.

The rate processes in disordered systems have a qualitatively different behaviour: For them, the relative fluctuations generally do not tend towards zero in the limit of large volumes. Although there is no universal asymptotic behaviour, the most typical situation is that for which the relative fluctuations tend towards constant values different from zero. In this case, the fluctuations are called intermittent, and they make a significant contribution to the average values of the concentrations: The average concentrations computed by taking the fluctuations into account are very different from the corresponding deterministic values, computed by neglecting the fluctuations. A less-typical behaviour is the one where the relative fluctuations diverge to infinity in the limit of large volumes; this case corresponds to fractal kinetics.

These features occur both for systems with static and for systems with dynamical disorder. A typical example is a first-order chemical reaction, $A \to$ Products, described by the kinetic equation

$$\mathrm{d}N/\mathrm{d}t = -kN. \tag{20.4}$$

For a system without environmental fluctuations, the dynamical behaviour of this process is trivial. For a system with environmental fluctuations, however, even though the evolution (20.4) is linear in the number of particles, N, this equation describes some non-linear coupling effects between the variable, N, and the rate coefficient, k, which is a random variable rather than a known number. If only environmental fluctuations are taken into account, and the sampling (chemical) fluctuations are neglected, the average number of molecules at time t, $N(t)$, can be evaluated by repeated integration of the differential (20.4) for different random trajectories of the rate coefficient, $k = k(t)$, and by taking an average over all possible trajectories $k = k(t)$,

$$\langle N(t) \rangle = \left\langle \exp\left[- \int_0^t k(t') \, \mathrm{d}t' \right] \right\rangle . \tag{20.5}$$

From (20.5) we see that not only the average value of the rate coefficient $k = k(t)$ contributes to the average value of the number of molecules, $\langle N(t) \rangle$, but rather all cumulants of the rate coefficient. This is true not only for dynamical disorder, where the rate coefficient is a random function of time, but also for static disorder, where the rate coefficient is a random number.

The evaluation of stochastic averages of the type in (20.5) is not a trivial problem, not even in cases of isolated reactions of first or second order. For simple reactions, analytic solutions are available in some cases, based on the method of characteristic functionals, or on the method of generalized cumulant expansion suggested by Lax [11,12] and Van Kampen [13] and expanded by others [14,15]. We outline only the main physical significance of the method of expanded cumulant expansion, which starts out from a general kinetic equation of the type

$$\mathrm{d}C(t)/\mathrm{d}t = \Phi\left[\mathbf{C}(t); \mathbf{k}(t)\right], \tag{20.6}$$

where $\Phi[\mathbf{C}(t);\ \mathbf{k}(t)]$ is generally a non-linear function of the composition vector $\mathbf{C}(t)$ and of the vector $\mathbf{k}(t)$ of the rate coefficients. By using the cumulant expansion technique, an infinite chain of evolution equations can be derived from (20.6). This chain of equations describes the relationships between the moments of the composition vectors and the various cumulants of the rate coefficients. In the case where the fluctuations of the rate coefficients are weak and have correlations that decay fast, an effective evolution equation for the average composition vector, $\langle C(t) \rangle$ can be derived:

$$\mathrm{d}\langle \mathbf{C}(t) \rangle/\mathrm{d}t = \Psi\left[\langle \mathbf{C}(t) \rangle\right], \tag{20.7}$$

where the effective (renormalized) vector of the reaction rates, $\Psi[\langle \mathbf{C}(t) \rangle]$ is generally different from the vector of the (bare) fluctuating reaction rates, $\Phi[\mathbf{C}(t); \mathbf{k}(t)]$. In this context, the cumulant expansion technique has a significance similar to the renormalization technique from quantum field theory: The influence of the environment on the reaction system is taken into account by replacing the vector of bare (fluctuating) reaction rates by a vector of dressed (renormalized) effective reaction rates [16]. Similar techniques are used for the study of wave propagation in random media and for the description of transport processes in disordered lattices [17].

Alternative techniques have been developed for the particular case of Markovian fluctuating rates. The main assumption is that the fluctuations of the rate coefficients can be described by a local evolution stochastic equation [16,18]:

$$\frac{\partial}{\partial t} P(\mathbf{k}; t) \equiv LP(\mathbf{k}; t), \tag{20.8}$$

where L is a linear integral or differential evolution operator of the Fokker–Plank or the Master equation type. The main idea is to introduce a joint probability distribution for the composition vector and for the rate coefficients, $\mathscr{P}(\mathbf{C}, \mathbf{k}; t)$. This joint probability distribution obeys the stochastic Liouville equation

$$\frac{\partial}{\partial t}\mathscr{P}(\mathbf{C}, \mathbf{k}; t) = L\mathscr{P}(\mathbf{C}, \mathbf{k}; t) - \nabla_{\mathbf{C}} \cdot [\Phi(\mathbf{C}; \mathbf{k})\, \mathscr{P}(\mathbf{C}, \mathbf{k}; t)]. \tag{20.9}$$

By solving this equation it is possible to evaluate the moments of the composition vector.

These techniques fail for strong environmental fluctuations. An interesting case of strong fluctuations is that where the rate coefficients obey Levy statistics. In the particular case of first-order processes, Levy fluctuations lead to stretched exponential integral kinetic equations of the Kohlrausch Williams–Watts (KWW) type:

$$\langle C(t)\rangle / \langle C(0)\rangle = \exp\left[-(\omega t)^{\alpha}\right], \tag{20.10}$$

where ω is a characteristic frequency and α is a dimensionless scaling exponent between zero and one, $1 > \alpha > 0$. The stretched exponential law is encountered not only in chemical kinetics but also in other chemical and physical rate processes occurring in disordered media. It was first proposed in 1864 by Kohlrausch to describe mechanical creep [19] and was later used to describe the dielectric relaxation in polymers [20] and for describing the failure data in reliability theory [21]. More recently, the KWW law has been used to fit the data on remnant magnetization in spin glasses, on the decay of luminescence in porous glasses, on the relaxation processes in viscoelasticity, on the reaction kinetics of bio-polymers [9], and on the dynamics of recombination kinetics in radiochemistry [6, 7]. Further applications include the description of the statistical distributions of open and closed times of ion channels in molecular bio-physics [22], and even the description of the survival function of cancer patients [23].

The ubiquity of the stretched exponential law has led to the idea that it should be generated by some kind of universal mechanism that is independent of the details of a given individual process. An argument in favour of this opinion is the close connection between the KWW law and the stable probability densities of the Levy type, which emerge as a result of the occurrence of a large number of independent random events described by individual probability densities with infinite moments [24,25]. Many attempts to search for such a universal mechanism of occurrence of the stretched exponential have been presented in the literature. A first attempt is a generalization of a mechanism of parallel relaxation, initially suggested by Forster [26] for the extinction of luminescence and extended by other authors [5]. A second model assumes a complex serial relaxation on a multi-level abstract structure, which emphasizes the role of hierarchically constrained dynamics [27]. A third model is a generalization of the defect-diffusion model of Shlesinger and Montroll [28]. All three models were carefully examined by Klafter and Shlesinger [29]; they showed that in spite of the different details of the three models, there is a universal common feature: the existence of a broad spectrum of relaxation rates described by a scale-invariant distribution. A complementary approach of the universal features of the stretched exponential is based on the powerful technique of fractional calculus and its connections with the theory of Fox functions [30].

A different approach to stretched exponential kinetics has been suggested by Huber [31]. Based on a careful examination of the models used for the description of the extinction of luminescence, he has derived a general relaxation function:

$$\langle C(I)\rangle / \langle C(0)\rangle = \exp\left[-\int_0^\infty \rho(\omega)\left[1-\exp(-\omega t)\right] \mathrm{d}\omega\right], \tag{20.11}$$

where $\rho(\omega)\mathrm{d}\omega$ is the average number of channels involved in the relaxation process and characterized by an individual relaxation rate between ω and $\omega + \mathrm{d}\omega$. The stretched exponential corresponds to a scaling law of the negative power law type

$$\rho(\omega)\,\mathrm{d}\omega \sim \omega^{-(1+\omega)}\mathrm{d}\omega, \tag{20.12}$$

which is consistent with the general ideas of self-similarity developed by Klafter and Shlesinger [29].

A number of generalizations of the Huber approach have been reported in the literature. It has been shown that Huber's equation is exact for a Poissonian distribution of independent channels [32]. Moreover, Huber's equation also holds beyond the validity range of Poissonian distribution: It emerges as a universal scaling law for a uniform random distribution of a large number of channels characterized by non-intermittent fluctuations [33,34]. Also, a second universal relaxation law has been identified that includes (20.11) as a particular case. For finite intermittent fluctuations, this equation predicts a crossover from a stretched exponential behaviour for moderately large times to a negative power law for very large times.

Equation (20.11) correspond to systems with static disorder. Similar equations have been derived for systems with dynamical disorder. The resulting equations have the same structure with the difference that the density of states is replaced by a functional density of states, and the integrals over the numbers of states are replaced by functional integrals [35]. The analysis of the asymptotic behaviour of these functional equations is complicated. However, a general pattern emerges, i.e., the stretched exponential relaxation function is stable; it is insensitive to the perturbations generated by the fluctuations of the numbers of channels.

Some progress has been made in the direction of applying the thermodynamic and stochastic theory of rate processes presented here to disordered systems. In some cases [35] it is possible to construct a stochastic potential with the properties the same as that for ordered systems discussed in Chaps. 2–11. A general set of fluctuation–dissipation relations has been derived that establishes a connection between the expression of the average kinetic curve,

$$\rho(\omega)\,\mathrm{d}\omega \sim \omega^{-(1+\omega)}\mathrm{d}\omega,$$

and the factorial moments,

$$F_\omega(t) = \langle N(N-1), \ldots, (N-m+1)\rangle\,(t)$$

of the number of molecules present in the system at time t [14,15]:

$$F_\omega(t) = [\langle C(t)\rangle / \langle C(0)\rangle]^\omega\,. \tag{20.13}$$

The problems are much more complicated, when, in addition to the non-linear coupling between the rate coefficients and the concentrations, the kinetics of the process is also non-linear. This problem can, however, be studied analytically if the concentration fluctuations are neglected.

Acknowledgement. This chapter follows a prior review of this subject [36], with editorial changes.

References

1. M.F. Shlesinger, Annu. Rev. Phys. Chem. **39**, 269–290 (1988)
2. R. Zwanzig, Acc. Chem. Res. **23**, 148–152 (1990)
3. A. Plonka, *Time-Dependent Reactivity of Species in Condensed Media* (Springer, Berlin, 1986)
4. A. Blumen, H. Schnörer, Angew. Chem. Int. Ed. **29**, 113–125 (1990)
5. A. Blumen, J. Klafter, G. Zumofen, In *Optical Spectroscopy of Glasses*, ed. by I. Zchokke (Reidel, Dordrecht, 1986) pp. 199–265
6. A. Plonka, *Time-Dependent Reactivity of Species in Condensed Media* (Springer, Berlin, 1986)
7. A. Plonka, Annu. Rep. Sect. C, R. Soc. Chem. **91**, 107–174 (1994)
8. R. Zwanzig, Acc. Chem. Res. **23**, 148–152 (1990)
9. A. Ansari, J. Berendzen, D. Braunstein, B.R. Cowen, H. Frauenfelder, et al. Biophys. Chem. **26**, 337–355 (1987)

10. M.O. Vlad, M.C. Mackey, J. Ross Phys. Rev. E. **50**, 798–821 (1994)
11. M. Lax, Rev. Mod. Phys. **38**, 359–379 (1966)
12. M. Lax, Rev. Mod. Phys. **38**, 541–566 (1966)
13. N.G. Van Kampen, Phys. Lett. A **76**, 104–106 (1980)
14. R. Zwanzig, J. Chem. Phys. **97**, 3587–3589 (1992)
15. M.O. Vlad, J. Ross, M.C. Mackey, J. Math. Phys. **37**, 803–835 (1996)
16. N.G. Van Kampen, *Stochastic Processes in Physics and Chemistry,* 2nd edn. (North-Holland, Amsterdam, 1992)
17. J.W. Haus, K.W. Kehr, Phys. Rep. **150**, 263–406 (1987)
18. R. Kubo, Adv. Chem. Phys. **15**, 101–127 (1969)
19. E.W. Montroll, J.T. Bendler, J. Stat. Phys. **34**, 129–162 (1984)
20. G. Williams, D.C. Watts, Trans. Faraday Soc. **66**, 80–85 (1970)
21. D.R. Cox, *Renewal Theory* (Chapman & Hall, London, 1962)
22. L.S. Liebovich, J. Stat. Phys. **70**, 329–337 (1993)
23. P.R.J. Burch, *The Biology of Cancer: A New Approach* (University Park Press, Baltimore, MD, 1976)
24. P. Lévy, *Théorie de L'Addition des Variables Aleatoires* (Villars, Paris, 1937)
25. B.V. Gnedenko, A.N. Kolmogorov, *Limit Distributions for Sums of Independent Random Variables* (Addison-Wesley, Reading, MA, 1954)
26. T. Förster, Z. Naturforsch. Teil A **4**, 321–342 (1949)
27. R.G. Palmer, D. Stein, E.S. Abrahams, P.W. Anderson, Phys. Rev. Lett. **53**, 958–961 (1984)
28. M.F. Shlesinger, E.W. Montroll, Proc. Natl. Acad. Sci. USA **81**, 1280–1283 (1984)
29. J. Klafter, M.F. Shlesinger, Proc. Natl. Acad. Sci. USA **83**, 848–851 (1986)
30. W.G. Glöckle, T.F. Nonnenmacher, Macromolecules **24**, 6426–6434 (1991)
31. D.L. Huber, Phys. Rev. B. **31**, 6070–6071 (1985)
32. M.O. Vlad, M.C. Mackey, J. Math. Phys. **36**, 1834–1853 (1995)
33. M.O. Vlad, M.C. Mackey, B. Schönfisch, Phys. Rev. **53**, 4703–4710 (1996)
34. M.O. Vlad, D.L. Huber, J. Ross, J. Chem. Phys. **106**, 4157–4167 (1997)
35. M.O. Vlad, R. Metzler, T.F. Nonnenmacher, M.C. Mackey, J. Math. Phys. **37**, 2279–2306 (1996)
36. J. Ross, M.O. Vlad, Annu. Rev. Phys. Chem. **50**, 51–78 (1999)

Index